PACIFIC SALMON FIELD GUIDE

WATERPROOF EDITION

SEAN GODWIN
MARTIN KRKOSEK
Joseph R. Tomelleri, Illustrator
Tavish Campbell, Photographer

LONE PINE

Lone Pine Publishing

Distributed by: Canada Book Distributors - Booklogic
www.canadabookdistributors.com
www.lonepinepublishing.com
Tel: 1-800-661-9017

Title: Pacific salmon field guide / Sean Godwin & Martin Krkosek.
Names: Godwin, Sean, author. | Krkosek, Martin, author.
Identifiers: Canadiana (print) 20220181039 | Canadiana (ebook) 20220181047 | ISBN 9781774511343 (softcover) | ISBN 9781774511350 (waterproof softcover) | ISBN 9781774511367 (PDF)
Subjects: LCSH: Pacific salmon—Identification. | LCSH: Pacific salmon. | LCGFT: Field guides. Classification: LCC QL638.S2 G63 2022 | DDC 597.5/6—dc23

Editorial: Erin McCloskey
Layout, Production & Design: Gregory Brown
Maps: Gregory Brown (except page 47 Coastal Cutthroat Trout Interagency Committee, 2019)
Cover Image Credits: Tavish Campbell
Image Credits: Tavish Campbell (except pages 42, 44, 50 Fernando Lessa)
Illustration Credits: © Joseph R. Tomelleri

PACIFIC SALMON FOUNDATION Proudly supported by the Pacific Salmon Foundation

We acknowledge the financial support of the Government of Canada.
Nous reconnaissons l'appui financier du gouvernement du Canada.

Funded by the Government of Canada
Financé par le gouvernement du Canada
Canada

Printed in China
PC: 38-1

CONTENTS

FOREWORD

If you are reading this book, you are likely a researcher, a naturalist or a fish enthusiast, and you likely have a personal library of favourite books in this genre. If so, make room for another on your shelf! I certainly will! I am nearing the end of my career and can easily identify the books that I love and have influenced my work. To be brief, there's *Return to the River* (R.L. Haig-Brown, 1941), *Salmon, Our Heritage* (C. Lyons, 1969), *Trout and Salmon of North America* (R.J. Behnke with illustrations by J.R. Tomelleri, 2002), *Behaviour and Ecology of Pacific Salmon and Trout* (T.P. Quinn, 2005) and most recently *Ocean Ecology of Pacific Salmon and Trout* (R.J. Beamish, Ed., 2018). These few books from hundreds in my library are not an adequate acknowledgement of the other authors, but I do so to emphasize the notability I place on *Pacific Salmon Field Guide*. This is the most comprehensive field guide on Pacific salmon and will accompany researchers and enthusiasts alike.

But it is not simply a field guide! In my opinion, it is a reference book rich in experience from skilled observers including clear description and names applied to the life stages of Pacific salmon, and with beautiful presentations of the rich diversity in Pacific salmon... most of which you will likely never see! Sean Godwin and Marty Krkosek have produced a wonderful resource that shares their great knowledge and in-depth appreciation of Pacific salmon, plus that of the many researchers that also contributed. The guide includes 58 original Tomelleri plates created for this book (69 plates in total) covering all seven species of Pacific salmon in North America and multiple life-stages within each species. For those unfamiliar with Joe Tomelleri's works, you will discover an artist with enormous talent who has hugely enhanced the value of this book by visualizing the important diversity in salmon. It is a tribute to nature and the intricate ties between salmon and their many habitats, but it is also an expression of an essential key to the future survival of Pacific salmon.

The rich diversity of salmon expresses the importance of different habitats across life stages and the adaptability of salmon over thousands of generations. But what we see today reflects their past. Today the numbers of salmon are greatly diminished, human impacts on salmon have been extensive, and now we have climate change. This field guide is an important reminder of this complexity and the essential need to ensure diverse habitats so salmon may adapt to a different but unknown future. I thank Sean and Marty for an important, timely and beautiful reminder!

When the authors approached the Pacific Salmon Foundation for financial support to produce a field guide for juvenile Pacific salmon, including art by Joe Tomelleri, I was immediately supportive. This book exceeds our expectations. Congratulations on an exceptional outcome that will benefit salmon and all who are interested in these charismatic species for years to come!

Dr. Brian Riddell is a science advisor for the Pacific Salmon Foundation, having recently "retired" after serving 10 years as President and CEO. His storied career spans more than four decades between the federal government and PSF, during which time he has co-authored Canada's Wild Salmon Policy, *advised on the Pacific Salmon Treaty with the United States and served as a Commissioner for the Pacific Salmon Commission (2010–2022).*

INDIGENOUS PERSPECTIVES

Nìt (hello) and welcome readers. Here I would like to honour and focus on the ethics, practices and principles that have arisen from thousands of years of co-existence and interdependence between *hoon* (meaning fish in Nisg̱a'a, but especially salmon) and Indigenous Peoples across the Pacific Northwest.

While I did not grow up in Nisg̱a'a territory, with my language, or alongside *hoon*, I have had the tremendous privilege to find my way home to these through my research and experiences as a fisheries scientist. Over the course of my doctorate, I spent time in many communities across three large home rivers for *hoon*—the Fraser, Skeena and Nass Rivers (home of the Nisg̱a'a). Here I share some of the words of those who were willing to share their time and knowledge with me during my studies. This is not written to represent the nature of all *hoon*–people relationships but reflects lessons I have come to learn through my work and time in community. Indigenous Peoples and knowledges are not a monolith. They share, however, important parallels when it comes to this identity-defining and cultural keystone species: *the salmon*. These concepts hold the power to improve how we collectively manage and interact with *hoon* today. At their core, they center on **taking only that which is given and taking only what one needs.**

As noted by a Secwepemc knowledge keeper "wild salmon are our relatives, and they're our most important food... talk[ing] about wild salmon as a resource to be exploited, as opposed to an intimate relative, are two very contradicting ways of relating to wild salmon. It goes against the Indigenous concept of all our relations." A St'át'imc Elder shared with me "We were raised to take what you need. If you take more than you need, then you give away what you don't need." A Nisg̱a'a elder simply and eloquently said "We have to share...We have to respect...We have to listen to nature." To ensure that not too many salmon are taken by one individual or for any one family, a Tsal'álh knowledge keeper noted "The first fish you catch, you let it go right? So, it goes. [That is] respect for the salmon." In the Xwísten community, a spiritual leader spoke of the First Salmon Ceremony that takes place each year upon the return of the first salmon across communities throughout the Pacific Northwest, stating "That's what the Salmon Ceremony is. Catching them and, as soon as you catch them from the river, you put in a little bit of tobacco and tell them—tell the Fraser—thank you for sending me that salmon that we need to nourish our bodies." He extended this gratitude to many, saying "Thank you Creator for the fish you sent up so that my people will not starve. And thank you for the river, and I pray that you cleanse our waters so that we can drink anywhere that we are at."

Hoon must be treated with respect if we expect our relationship with them to continue. The connection between the health of *hoon*, the health of people and the health of the river was not lost on any knowledge keepers I had the privilege of listening to. As you, reader, put this guidebook to use, it is my hope that these principles and practices will permeate your own thinking. As you move through waterways in search and study of salmon, think about how you can show respect and uphold your responsibilities to all our relations.

Dr. Andrea Reid wrote this from the territories of the xʷməθkʷəy̓əm *(Musqueam),* Sḵwx̱wú7mesh *(Squamish) and* səl̓ílwətaʔɬ *(Tsleil-Waututh) in what is now known as Vancouver, British Columbia. Dr. Reid is an Assistant Professor at the Institute for the Oceans and Fisheries at The University of British Columbia, and she is the Principal Investigator with the Centre for Indigenous Fisheries. She is a citizen of the Nisg̱a'a Nation; her paternal side from the Nisg̱a'a Village of Gingolx and following the matrilineal line to the* Gisḵ'aast *(killerwhale) clan and to* Wilp *(house)* Daax̱an. *Maternally, she is of Irish descent and was raised on Epekwitk (Prince Edward Island) with her two older brothers by their mother. She is a daughter, sister, auntie and lover of all things aquatic.*

ACKNOWLEDGEMENTS

A book like this may not be very long, but it takes an army of people to make possible.

By the time this book hits the shelves, it will have been seven years since we (the authors) first conceived of it over beers in a Vancouver pub. At the time, we envisioned a short field guide for only juvenile Pacific salmon. We had spent the past several years researching post-smolts in coastal British Columbia, lamenting the lack of identification resources for that life stage. Little did we know what this modest idea would eventually grow into.

First, we cannot express enough thanks to Joseph Tomelleri, who drew all the beautiful illustrations in this book. Joe is peerless in his talent and meticulous in his detail. Beyond his artistic abilities, Joe is as knowledgeable and passionate about fish as anyone we have met. We count our lucky stars that he took on this project.

The illustrations in this book would not have been possible without the generous financial backing of the Pacific Salmon Foundation. We are particularly thankful to Brian Riddell who was supportive throughout the entire process, and wrote the wonderful (and kind) foreward for this book.

We are grateful to Tavish Campbell, who contributed the stunning photography showcased in this book. It was a delight to select photos from his magnificent portfolio—we wish we could have included more. We also thank Fernando Lessa for the three beautiful steelhead and coastal cutthroat photos he provided to fill in the gaps.

It is scary to think what this book would have looked like without the hard work (and patience) of our editor, Erin McCloskey. We knew from day one that this book was in good hands with the excellent team at Lone Pine Publishing, including Shane Kennedy, who believed in this project from the outset, and Greg Brown, who brought the book to life through its design.

We thank Andrea Jane Reid for writing the excellent piece on salmon and Indigenous Peoples for the beginning of this book and for highlighting such an important subject in her writing. We are indebted to the Indigenous Peoples of this coast, who have stewarded Pacific salmon for time immemorial and who have watched colonization result in widespread salmon declines. We—the authors and settlers generally—have much to learn from Indigenous perspectives on respect and responsibility towards salmon and the other wildlife with which we share land and water. We must also recognize and respect the history of the places in which we live and work; to that end, we acknowledge and are thankful for the opportunity to have written this book from the lands and waters of the unceded traditional territories of the Kwakiutl, Musgamagw Dzawada'enuxw, 'Namgis, Musqueam, Squamish and Tsleil-Waututh First Nations.

We relied on the expertise of a huge number of people to make sure this book was as accurate and comprehensive as possible. James Losee is one of the leading scientific authorities on coastal cutthroat and was very generous with his time to make sure we represented this underappreciated species well. Patrick Trotter, Riley Freeman and Kitty Griswold were also key sources of inspiration and information for the coastal cutthroat parts of this book. Lauren Portner, Dylan Smyth and Mirko Diaz were major contributors to the indentification sections of this book, enduring countless back-and-forth dialogues on the particulars of marine-phase juvenile salmon. Colin Bailey was a vital resource for juvenile steelhead and cutthroat.

We are also enormously thankful for the additional expert feedback from Jason Assonitis, Brendan Connors, Julie Charbonneau, Katie Davidson, Cameron Freshwater, Nathan Furey, Julian Gan, Samantha James and Stephen Johnston.

Many of the specimens illustrated for this book were collected alongside or by researchers and practitioners other than ourselves. This book could not have been completed without these wonderful people. Specifically, we thank Fiona Francis and Debora Obrist (for their help collecting many juveniles); Shane Kalyn, Matthew Townsend and the DFO crew in Mission (several fry and smolts); James Losee and the WDFW coastal cutthroat crew (several coastal cutthroat); Marcus Boucher, Adrian Clarke and Dusty White (sockeye fry); Jason Brandt, Greg Huchko, Holly Huchko, Bruce McIntosh and David Noakes (steelhead fry and parr); Collin Middleton and InStream Fisheries Research (steelhead and coastal cutthroat smolts); Cheryl Morgan (steelhead post-smolt); Katherine Maslenikov (steelhead ocean adult); Michael Arbeider, Michael Graham and Adam Kurkiewicz (female Chinook spawner); and Mirko Diaz and Hamid Seshadri (many spawners).

We are so very grateful to our friends and family who supported us while writing this book. For MK, the countless adventures mucking about in creeks, lakes, boats and beaches, as child and adult, have been an inspiration—thank you in particular to MJ, Fife, Anika and Katia. Sean acknowledges his late grandfather, who instilled in him a love for salmon and the ocean at an early age, and the wonderful people of Salmon Coast Field Station (including MK), who introduced him to the joy of field biology.

Finally, we thank you, the reader, for your interest in Pacific salmon. We hope this book helps foster in you an appreciation for these special fishes and a desire to help protect them.

INTRODUCTION

Pacific salmon have been described in many ways—a keystone species, the lifeblood of this coast and an icon of the Pacific Northwest. Migrating thousands of kilometres from the Pacific Ocean to the mountain streams where they were born, salmon bring a flood of nutrients that supports an astounding abundance and diversity of life. Aquatic insects, songbirds, mosses, salmonberry, Sitka spruce, eagles, seals, bears and orcas—the health and survival of countless species are tied to salmon. On Kodiak Island, a grizzly bear can consume 30 salmon per day during a few short weeks of spawning, eating more than 3000 kg (or 6500 lbs) of fish during that time. Marine nitrogen from salmon can be found in the topmost needles of trees lining coastal streams and estuaries. Songbirds are more abundant around rivers with more salmon. Stream fishes, including young salmon, feed on the insects attracted to the salmon carcasses spread about the forest by bears. Fewer endangered southern resident orcas die in years of abundant Chinook salmon in the Salish Sea. Salmon underpin this entire food web.

Pacific salmon are spectacularly diverse. Their colouration, stripes and spotting patterns change over their life cycle and differ dramatically among populations and species. So too, their life histories differ (such as migrating long distances or not at all, staying years at sea or maturing young and sneaking mating opportunities). Among populations, there are many adaptations specific to the features of the local environment, such as stream temperature or the visual field of the open Pacific Ocean. Some rivers have mixtures of anadromous steelhead salmon and non-migratory rainbow trout, which are two forms of the same species that interbreed, and siblings vary in their decisions to migrate or not. Through time, there are enormous cycles of abundance that range from the famous four-year cycle of the Adams River sockeye to the unusual abundance of pink salmon that occurs only in odd years in one place and in even years someplace else. All this diversity within and among Pacific salmon populations and species somehow evolved and coexists on the northwest coast of North America.

This field guide to Pacific salmon covers the seven anadromous forms of species in the genus *Oncorhynchus* that are distributed along the northwest coast of North America: pink salmon, chum salmon, sockeye salmon, coho salmon, Chinook salmon, steelhead trout and coastal cutthroat trout. For convenience, we use the term "Pacific salmon" throughout this book to encompass these seven species. We have not included masu salmon, which is a Pacific salmon but only occur in Asia. The term "anadromous" refers to a life cycle in which a fish is born in freshwater, matures in the ocean and returns to freshwater to spawn. Many species of Pacific salmon have populations that are not anadromous because migration barriers, such as waterfalls and dams, prevent them from migrating to sea. Kokanee and rainbow trout are examples of non-anadromous fish that have anadromous counterparts of sockeye and steelhead with overlapping ranges and with which they interbreed. The genus *Onchorynchus* falls within the family Salmonidae, which includes other common fishes such as Atlantic salmon and some trout (*Salmo*), char (*Salvelinus*), whitefishes (*Coregonus, Prosopium and Stenodus*) and grayling (*Thymallus*). Species in these genera that overlap in distribution with Pacific

Sunrise over Okisollo Channel, BC, a major migration corridor for Pacific salmon

salmon include Arctic char (*Salvelinus alpinus*), Dolly Varden (*Salvelinus malma*), bull trout (*Salvelinus confluentus*), Arctic grayling (*Thymallus arcticus*), many whitefish species, introduced brown trout (*Salmo trutta*) and occasionally Atlantic salmon (*Salmo salar*) when they escape from salmon farms. Of the Pacific salmon, pink, chum and sockeye are the most abundant, steelhead and coastal cutthroat are the least and coho and Chinook fall in the middle. Chinook have the largest body size, whereas pink are the smallest, aside from non-anadromous populations.

BEAUFORT SEA
ALASKA
CANADA
GULF OF ALASKA
PACIFIC OCEAN
UNITED STATES
RIVERS
1 Nushagak River
2 Yukon River
3 Kenai River
4 MacKenzie River
5 Fraser River
6 Chilcotin River
7 Thompson River
8 Adams River
9 Stillaguamish River
10 Columbia River
11 Williamette River
12 North Upqua River
13 Sacramento River
14 San Jaquin River
15 Kern River
16 Snake River
LAKES
17 Sakinaw Lake
18 Cultus Lake
19 Kootenay Lake
20 Ozette Lake
21 Upper Klamath Lake

SALMON AND PEOPLE

Throughout their range, Pacific salmon have been central to the culture and food security of Indigenous Peoples. Far inland, groups such as the Tr'ondëk Hwëch'in, Tk'emlúps te Secwépemc and Kutenai harvested salmon thousands of kilometres up the Yukon, Fraser and Columbia watersheds. Coastal First Nations communities, such as the Kwakwaka'wakw and Nuu-chah-nulth, harvested pink and chum salmon using weirs in the estuaries of coastal streams. Humans have thus coevolved with salmon and have long been part of the ecosystem that salmon support. Pacific salmon were vital to European settler communities, providing food security and an economic resource base upon which coastal fishing communities grew and thrived. In the early to mid-1900s, the Pacific coast of North America was dotted with canneries, hotels, pubs, schools and float house communities along the estuaries and bays near productive salmon rivers, many now abandoned relics fading away into the wilderness alongside the middens of the peoples before them.

As vital as salmon have been to the Indigenous and settler peoples of the Pacific Northwest, modern society has not reciprocated in kind. Rather, the conservation status of Pacific salmon is overwhelmingly linked to the multitude

A group of sockeye nearing the end of their journey in the Adams River, BC

of anthropogenic impacts on salmon and their habitats. The legacy of overfishing by settler communities on what had seemed to be an inexhaustible resource, along with damage to their freshwater habitats from dams, pollution, logging and water diversion, have severely reduced salmon populations in the southern portions of their range. That history now combined with climate change has produced an ever-growing list of endangered species designations of once-abundant runs, species recovery working groups, fishery restrictions, conflict over fishery access, conflict over water, rivers too warm for salmon to survive and rivers that no longer reach the sea.

Meanwhile, in the north, salmon are expanding their range and are more productive than ever in recorded history, including the highest catches on record for the Bristol Bay sockeye salmon fishery—the largest salmon fishery in the world and a flagship of sustainable fisheries management. In Bristol Bay and farther north there are few people, no cities and less logging, damming and pollution. Climate change may make these northern habitats more productive for salmon. It is a lesson that if you leave the natural diversity of salmon and their habitats intact, it is possible for the fish and the fisheries to thrive. Despite the conservation concerns and fishery restrictions that we are accustomed to hearing about salmon from California through much of British Columbia, Pacific salmon overall are more abundant than they have been in many decades. That production, however, is in the north and increasingly slipping away from the backyards, rivers and coastal seas that surround Vancouver, Seattle, Portland and San Francisco.

Restoration of habitats has helped salmon. Removing dams or easing migration barriers, protecting forests along rivers and streams and leaving water in the streams are obviously essential for salmon to survive. So, too, are restricting harvests, protecting small at-risk populations and avoiding disease spread from hatcheries and fish farms. Restoring and conserving salmon involves managing people more so than fish, for what wild salmon need most is relief from human impacts. Even our most well-intentioned efforts, like building hatcheries to help endangered runs recover, can go awry; populations can become dependent on hatcheries for their survival and lose the evolutionary adaptations they had to survive on their own. We must recognize that recovery is not a foregone conclusion even if we do all the right things. By counting salmon scales in the sediments of lakes, we know that salmon have naturally cycled through decades and centuries of low abundance as the ecosystem rebalances, long before anthropogenic climate change.

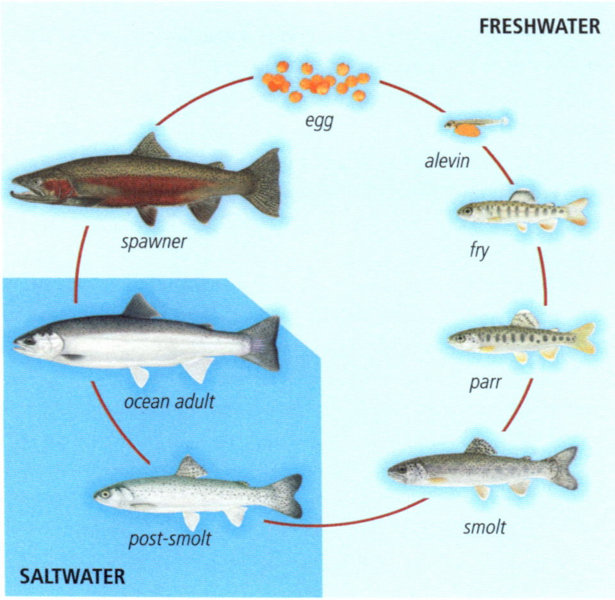

SALMON LIFE CYCLE

Pacific salmon are famous for their migrations between the Pacific Ocean and the rivers and streams of western North America. They spawn in the gravel of streams, rivers, lake beaches and even intertidal areas where there is a steady flow of clean freshwater to oxygenate the eggs. In spring, eggs hatch into alevins that rear in the interstitial spaces of their gravel nests (called redds) before emerging as fry. The eggs will have incubated from only several weeks to several months depending on the local environmental conditions and spawning period. Salmon naturally have low prospects of survival—each female produces thousands of eggs because few offspring will survive to spawn.

Pacific salmon fry are a point of departure among numerous life histories that vary among species, populations and even siblings. Pink and chum fry migrate to sea immediately after gravel emergence, where they grow rapidly but have poor survival rates, whereas other species rear in streams, rivers and lakes for one or more years before migrating to sea as smolts. However, Chinook salmon have stream-type and ocean-type forms, the latter of which also migrate to sea as fry. Steelhead salmon is the same species as rainbow trout; they can interbreed, and siblings can be anadromous (migrate to sea) or remain resident in streams, rivers and lakes for their entire life. Coastal cutthroat trout also vary in their migrations to sea, but those that go remain in the bays and estuaries nearby their natal rivers. Most steelhead and coastal cutthroat (and some sockeye, coho and Chinook) will spend one year or more in freshwater, at which point they are called parr rather than fry.

Once at sea, most juvenile salmon commence their coastal migration to their offshore feeding grounds in the open ocean, during which time they are called post-smolts. Salmon post-smolts and, eventually, adults enjoy an expanded and more productive variety of invertebrates and smaller fishes upon which to feed; however, they must contend with more predators, such as larger marine fishes, seabirds and marine mammals. Consequently, salmon species, populations and even siblings differ in their decisions to stay at sea longer to grow at the cost of exposing themselves to marine predators. Some males of several species return to rivers early, and with small body sizes and subdued colours they sneak in and surreptitiously fertilize eggs rather than risk the predators at sea.

After spending up to five years at sea, depending on the species, Pacific salmon migrate back to freshwater to spawn. Most Pacific salmon die after spawning. Millions of salmon carcasses feed wildlife and fertilize the landscape with nutrients from the sea. Steelhead and coastal cutthroat, on the other hand, can survive spawning and return to the ocean, migrating between the ocean to feed and freshwater to spawn multiple times over their lives. Some species,

Pacific salmon feed coastal wildlife, like this black bear on Princess Royal Island, BC.

An adult coho about to start its upstream ascent to spawn in the Cowichan River, BC

such as sockeye, often spawn in the same stretch of creek in which they were born, even in enormous watersheds like the Fraser River. They have an internal magnetic compass that helps them navigate the ocean; for example, whether salmon return to the Fraser River via the inside or outside of Vancouver Island depends on small variations in the Earth's magnetic field. Other species, like pink salmon, instead can return to one of many rivers in a coastal region where they were born.

Spawning runs of Pacific salmon occur between California and Alaska and are expanding into the Arctic. Pink, chum and coho are the most common salmon in coastal rivers and streams, while Chinook and a couple chum runs migrate far inland in large river systems. Sockeye are associated with lake habitats for rearing juveniles, but adults spawn in a variety of habitats, including lake beaches. Steelhead are found throughout coastal rivers and streams, whereas coastal cutthroat typically spawn in smaller tributaries and headwaters.

HOW TO USE THIS FIELD GUIDE

This field guide is an identification resource that includes natural history, distribution and identification of Pacific salmon, as well as information on observing these fish in the wild. We organized the identification portions of this field guide into five life stages: fry, smolt, post-smolt, ocean adult and spawner. The fry section includes parr descriptions for steelhead and coastal cutthroat. We did not include alevin (the stage immediately after hatching) because they are rarely seen and mostly indistinguishable right out of the egg. Once they are differentiable, they resemble fry.

Specialized terminology is explained in a glossary of definitions in the back of the book. The identification sections, by necessity, have scientific terms for body parts, which are detailed in the following illustrations. We do not describe every body part of the fish for each life stage, only those that are useful for differentiating between species. When we discuss fish length, we are describing the distance from the tip of the snout to the fork of the caudal fin (fork length). We focus on live fish identification in this book but include characteristics that can only be assessed from dead fish (the numbers provided in this section are what are usually observed but counts outside of these ranges do occur).

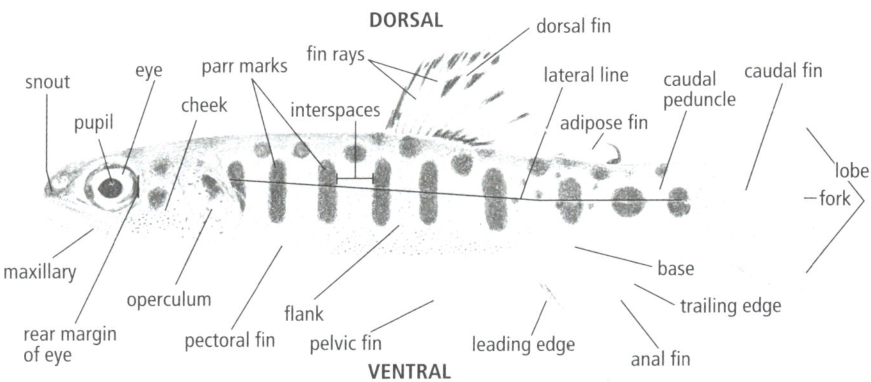

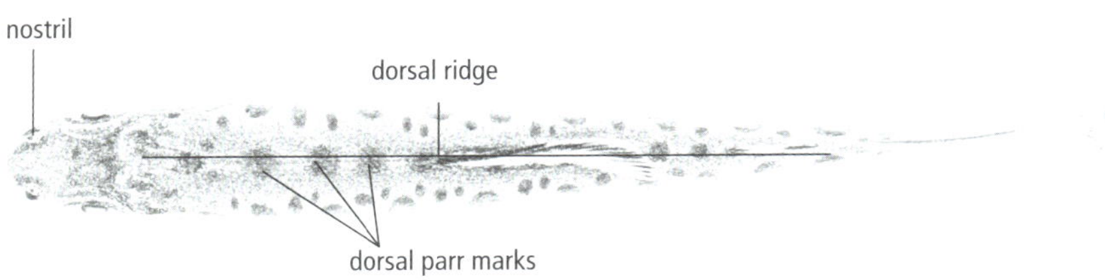

Salmon species can be difficult to differentiate. For most species and life stages, there is rarely a single distinguishing characteristic; instead, we often consider several characteristics as well as the context (such as time of year or habitat). We designed the identification sections in this book accordingly, by providing a summary table of key features across species for each life stage, as well as descriptions of behaviours and habitat. We include short field impression statements at the start of each life stage account to convey the look, demeanor and/or personality of the fish, informed by many biologists who work closely with them. For the juvenile life stages, we include top-down illustrations because this is often how fish are observed in the wild and because the dorsal surface can have many useful features for identifying young salmon. Finally, the identification sections include summaries that describe how to differentiate similar species.

We urge readers to keep live salmon in water when attempting to identify them. Salmon are not well equipped to be handled out of water, especially juveniles, which are fragile, and spawners, which have limited energy reserves. There are many simple ways to keep juvenile salmon in water while identifying them, such as with "fish viewers" or "photariums" that are specifically designed for this purpose, but water-filled buckets, Ziploc® bags and small aquariums are also great options for observing young salmon. For spawners, identification can usually be done quickly without removing them from the stream at all. Luckily, salmon are usually more straightforward to identify when in water, as

A young Theo Campbell-Hyatt holds a sockeye salmon in the Discovery Passage, BC.

Estuaries, like that of the Khutze River, are key habitat for Pacific salmon.

their colours and markings are easier to discern, and their fins stay splayed rather than collapsing. Magnifying glasses or hand lenses can be helpful for identifying juvenile salmon.

The scientific illustrations are from real specimens that were representative of their species and life stage. Joseph Tomelleri, who drew all the brilliant illustrations in this book, painstakingly measured every part of each fish, counting every fin ray and row of scales while layering colour upon colour. It is important to remember, however, that despite their incredible accuracy, these illustrations can only depict a single individual. Salmon are famous for showing incredible variation among individuals and populations. For example, northern Chinook smolts often have parr marks (the dark vertical markings on the flanks of fry and parr) that are more like southern coho than southern Chinook, and the dorsal spotting of sockeye post-smolts ranges from none to extensive, depending on the population. In this book, we attempt to describe this variation and use words like "usually" or "sometimes" to describe the frequency of occurrence in certain characteristics. This variation underscores the importance of weighing multiple characteristics when identifying salmon.

If you still can't identify a salmon even after careful assessment, that's okay! Some individuals are particularly difficult to identify and can only be accurately distinguished via genetic techniques, especially for certain species and life stages (Chinook and coho smolts or steelhead and coastal cutthroat fry, for instance). Our hope is that this book gives you the best possible chance at identifying these magnificent fish, while highlighting the beauty and diversity of Pacific salmon across all life stages.

THE SPECIES

Pink Salmon

(Oncorhynchus gorbuscha)

Humpback Salmon, Humpies

Pink salmon are the smallest but most abundant Pacific salmon. They are known as "humpback salmon" or just "humpies" in reference to the large, humped back of spawning males. Pink salmon have the shortest and most specialized life history of Pacific salmon—all humpies migrate straight to sea after hatching and spawn at two years of age. They primarily eat zooplankton and krill. Their abundance makes them the main source of salmon biomass that returns to coastal ecosystems most years.

Their two-year life cycle results in two distinct populations that inhabit each river: one that spawns in even years and another that spawns in odd years. One of these two population is often much greater than the other, resulting in alternating years of high and low abundance. On their spawning migration, most pink salmon return to their natal river, but many also return from the ocean to different rivers nearby or even distant from where they were born. Consequently, runs that are thousands of kilometres apart in the same year are more closely related than runs in the same river in opposite years.

The abundance of pink salmon provided food security to Indigenous Peoples in the Pacific Northwest. Pink salmon helped establish settler communities and their commercial fisheries and canneries, which processed, preserved and distributed worldwide the catch from the Pacific Northwest. Most of those canneries are now gone, and while commercial fisheries for pink salmon remain restricted in the southern part of their range, they are thriving in the north. The small size and low fat content of pink salmon make them a less prized culinary salmon, but they are delicious cooked fresh over a smoky cedar and alder campfire.

DISTRIBUTION

Pink salmon spawn mostly in the lower reaches of small coastal rivers or tributaries, and even in estuarine intertidal areas, from Washington to British Columbia and Alaska and over to Russia and parts of Asia. Their spawning distribution in rivers and streams tends to overlap that of chum but with spawning occurring earlier in the year. In the last several decades their distribution has expanded in the Arctic, including the McKenzie River, and contracted in California and Oregon. Pink salmon have been introduced and established in the Great Lakes as well as in Europe, where they are expanding their range.

LIFE CYCLE

Spawning occurs in late summer or early fall. After eggs hatch in late winter, the fry migrate through estuaries and into the ocean in early spring at only a few months of age (most other salmon remain in lakes, rivers or streams for at least a year). Although juvenile pink salmon are small when they enter the marine environment, they can be plentiful in protected nearshore habitats between the seashore and kelp beds, where they feed on zooplankton and grow in body size as they attempt to avoid predators. Once they start gaining scales, juvenile pinks move into deeper water beyond the kelp beds. During this early marine phase, they are heavily predated by seabirds and other fishes, including larger salmon, particularly coho.

In the summer of their first year, pinks migrate offshore to the Gulf of Alaska and Bering Sea, where they feed primarily on krill, other large zooplankton and small fishes. The winter at sea is marked by low food availability and high mortality. Pink salmon return to coastal marine waters in the summer of their second year and are about 3000 to 4000 times larger by mass than when they first migrated to sea as fry. In coastal marine waters, adult pink salmon are heavily predated by seals and sea lions, and even Pacific white-sided dolphins have been observed corralling shoals of pink salmon and feasting on them in narrow bays. Pink salmon aggregate in estuaries during late summer and early fall before migrating upstream to spawn in the lower reaches of rivers or sometimes estuaries in response to high rainfall and increased river flows.

As they prepare for spawning, pink salmon develop their spawning colours and body shape, which involves losing the silvery streamlined form of the marine phase. Males develop a pronounced hump between their head and dorsal fin—a sexual trait selected by females—as well as hooked upper jaws that they use in battles against other males as they compete for mating opportunities. The males darken above the lateral line becoming a mixture of dark green and blue, with localized reddening around the gill plate, lateral line and tail. Females develop similar colouration but do not develop hooked upper jaws. Spawning extends from August to mid-fall, depending on location, during which time there is high mortality from terrestrial predators such as bears.

A school of juvenile pink and chum caught in Johnstone Strait, BC

Pinks are the most abundant of all Pacific salmon (Campbell River, BC).

OBSERVING IN THE WILD

Tiny freshwater pink salmon fry are difficult to see as they migrate to the ocean. Upon reaching estuaries and marine waters, the surface activity of these small fish is easily observed in calm conditions—look for dimpling and occasional leaping. Adult pink salmon in marine waters are silvery and can be distinguished by their leap, which typically involves a high arch and a floppy sideward landing. On their spawning grounds, the striking colouration of both males and females makes them highly visible, the males even more so with their large hump. Large piles of odorous rotting carcasses on streambanks can provide clues that pink salmon are spawning, but in such situations be mindful of bears.

CONSERVATION

Pink salmon are not considered to be of particularly high conservation concern because they have an overall high abundance and are thought to be the most resilient of the Pacific salmon. Although pink salmon abundance declined during the era of local canneries and over-harvesting in the early to mid-1900s, numbers in the North Pacific Ocean have since recovered and are near record highs. Contractions of the species range in California and Oregon may reflect anthropogenic effects, including climate change, migration barriers and changes to freshwater habitats. However, historical records in these areas may indicate adults straying from other spawning areas. The main distribution of self-sustaining pink salmon populations begins in Washington and is expanding in the Arctic, into the Mackenzie River and other Arctic rivers.

Chum Salmon

(Oncorhynchus keta)

Dog Salmon, Keta Salmon, Silverbrite Salmon

Chum salmon are the second most abundant salmon in the North Pacific Ocean after pink salmon. Their name "chum" is the Chinook word for "striped," describing their tiger-like stripes during spawning. The "dog" nickname reflects the large jaws and teeth that males develop to compete for mating opportunities. Chum are sometimes labelled "silverbrites" in stores to make them more marketable. Like pink salmon and ocean-type Chinook, chum do not rear in freshwater environments, but rather migrate to sea shortly after emerging from their redds. Chum salmon have the greatest body depth of all Pacific salmon, and only Chinook salmon have a larger body length.

Before returning to freshwater to spawn, most chum salmon spend three winters at sea, with more than 90% three-winter fish in some populations, resulting in four-year cycles varying between high and low abundance. The proportion of fish that are two-winter and four-winter varies among populations, with more older fish in the north and more younger fish in the south, but these are rarely more than 50%. They typically spawn in smaller coastal systems, but amazingly also have two major river migrations that are among the largest of all Pacific salmon—more than 3200 km (more than 2000 mi) in the Yukon and Amur rivers, where there are distinct summer and fall runs. At sea, adult chum salmon eat smaller fish and invertebrates, including squid and some jellyfish, but like other salmon, they do not actively feed during spawning or during their upstream river migration.

Indigenous Peoples of the Pacific Northwest have relied upon the abundance of chum salmon for food security. Chum similarly supported the development of settler fishing communities and canneries in the late 1800s and early 1900s, which resulted in overexploitation and population declines. Chum salmon are produced by hatcheries for ocean ranching more than any other Pacific salmon because they require minimal freshwater rearing and grow large from fry to adult, but most of that hatchery production comes from Asia. Their eggs and fillets are valued for sushi in Japan, but their low fat content makes them less valuable than sockeye, coho and Chinook in North America. However, cold-smoked chum salmon are particularly delicious.

DISTRIBUTION

Chum salmon have the broadest distribution of Pacific salmon, ranging from California to the Mackenzie River in the Arctic. Aside from the anomalous long-distance migrations in the Yukon and Amur rivers, chum salmon typically spawn in the lower reaches of coastal rivers and streams in similar locations to pink salmon, but later in the fall. The southern range of chum salmon has contracted, particularly in California, where returns are now inconsistent. However, in the North Pacific Ocean, their overall abundance has been increasing through natural population growth as well as hatchery production in Asia.

LIFE CYCLE

Spawning occurs in mid-fall except for the unique summer runs of the Yukon and Amur rivers. Chum salmon do not rear in freshwater environments; instead, they emerge from their redds and immediately migrate to sea. They enter estuaries to adjust to the marine environment. During their early marine period they are often still called fry rather than smolts because they have a small body size in comparison to co-occurring coho, sockeye and most Chinook, which are normally one year older. After entering the ocean, chum salmon grow rapidly, feeding primarily on zooplankton. They are heavily predated by seabirds and other fishes, including larger salmon smolts and sea-run trout.

Chum post-smolts from the Pacific Northwest migrate northwest to the Gulf of Alaska, where they typically spend between two and four winters; however, a few chum salmon spend only one winter or as many as five winters at sea. In the North Pacific Ocean, their distribution can overlap with stocks of chum salmon from Russia and Asia. While at sea, chum salmon feed on pteropods, krill, amphipods, smaller fishes and squid. Upon their return migration, chum salmon have grown around 12,000 times by weight, with males usually larger than females.

On their spawning migration, chum salmon are predated by marine mammals such as orcas, seals and sea lions, and are also caught in Indigenous, commercial and recreational fisheries. Upon entering rivers, chum salmon lose their silvery stream-lined appearance and develop their spawning characteristics, including the purple tiger stripes, humped backs and hooked

Chum post-smolts caught in Okisollo Channel, BC, alongside pink, coho and sockeye

The carcasses of chum salmon provide vital nutrients to the ecosystem (Mussel River, BC).

upper jaws of the males. Females also take on a striping pattern. They can degrade quickly after entering streams and rivers, developing conspicuous fungal infections. The large numbers of spawning chum salmon are a major source of nourishment for many coastal species including invertebrates, fishes, birds, wolves, bears and even plants and trees.

OBSERVING IN THE WILD

Juvenile chum migrate to sea shortly after emerging from their redds, and mostly at night, making them difficult to observe in freshwater environments. In coastal marine waters, juvenile chum salmon can be found intermingling with pink and sockeye salmon, often in surface waters near shore but outside kelp beds. As juveniles and adults, they are distinguishable by their leaps, which often occur as a series of shallow leaps in quick succession that form a u-shaped pattern. Large, colorful and abundant spawning adults are conspicuous in their typically smaller river systems.

CONSERVATION

There are two imperilled populations of chum salmon in the southern portion of their range—the Columbia River and the Hood Canal summer-run populations are listed as threatened in the United States under the Endangered Species Act. However, overall species abundance is high in the North Pacific Ocean and has been increasing in recent decades in association with favourable ocean conditions and high hatchery production in Asia.

Sockeye Salmon

(Oncorhynchus nerka)

Red Salmon, Blueback Salmon

Of the Pacific salmon (but perhaps not sea-run trout), sockeye have the most diverse life history and, because they are abundant and have delicious red, oily flesh, are the most commercially valuable. They rank third for abundance behind pink and chum and have the second-smallest body size, with pinks being smallest. The name "sockeye" does not refer to their eyes but is an anglicized version of the Salish name for the fish—variations on *scəqay* among the Indigenous Peoples of the Salish Sea and lower Fraser River area of British Columbia. Sockeye are also called "red salmon" for their deep red fillets and spawning colouration. Non-anadromous, or lake-resident, forms of sockeye are called "kokanee" meaning "red fish" in the language of the Kutenai people of British Columbia, where the fish occurs in Kootenay Lake.

Sockeye salmon are famous for their variety of spawning habitats, their fidelity to natal spawning sites and their variety of body forms that are locally adapted to these habitats. They usually spawn in tributaries of lakes, including the mainstems of large rivers, and in tiny creeks but also along upwelling beaches of the lakes themselves. Females prefer to breed with large males with humped backs—these features become exaggerated in lake-dwelling populations where the fish can easily escape bear predation. Creek spawners, however, can suffer intense predation from terrestrial predators, which keeps the evolution of large-humped bodies in check. Juvenile sockeye typically rear in lakes, feeding on invertebrates, before they migrate to sea at one to three years old. They spend one to three years at sea before returning to spawn.

Kokanee often occur in river systems that have a migration barrier to sea—anthropogenic or natural. These fish may have originated from sockeye runs that survived after a dam or other natural migration barrier formed. However, kokanee overlap with sea-going sockeye, sometimes spawning in proximity but not interbreeding. Sockeye and kokanee remain closely related genetically, therefore biologists consider them an example of evolution in action where one species has formed two populations that have begun diverging into separate species. On spawning grounds, kokanee are miniature vivid versions of sockeye, and they have a similar body size with sockeye jacks, which are the youngest and smallest spawners. Jacks obtain mating opportunities not by competing for mates but by waiting for a female to commence spawning with a larger male and then sneaking in to fertilize some of the eggs. Sockeye jacks are rare compared to coho and Chinook, but the sneaky mating strategy remains the same.

Sockeye have been important to the culture and food security of Indigenous peoples throughout the fish's range. Sockeye are harvested by commercial and traditional methods. They were and are often smoked over cedar fires and sun dried. Today, sockeye salmon in Bristol Bay, Alaska, support the largest and most commercially valuable salmon fishery in the world. It is an example of a sustainable fishery, where the diversity of the fish and the integrity of its habitat remain intact and help maintain stable and productive returns and harvests. In contrast, southern runs have declined, and fisheries have become restricted or closed for a variety of reasons, with climate change, habitat alteration and a loss of diversity among many concerns.

DISTRIBUTION

Of Pacific salmon, sockeye have the narrowest latitudinal spawning range. They spawn in river systems from the Columbia River in Oregon to Bristol Bay and the Aleutian Islands over to eastern Russia. Their distribution is also linked to lakes, which sockeye rely on for juvenile rearing. Most of the interior populations in the Columbia have been lost to dams. Sockeye become much more rare north of Bristol Bay (for example, there are few runs in the Yukon River) and intermittent in the Arctic because they are reliant on glaciated landscapes with lakes. Kokanee are found throughout this range, often but not always associated with migration barriers. The marine distribution of sockeye is coastal during the first summer at sea but then offshore in the Gulf of Alaska and Bering Sea until adult sockeye make their return migration to spawn.

LIFE CYCLE

The timing of sockeye migrating back to spawn ranges from early summer to mid-fall. Runs are typically earlier in the north and later in the south, but even within a river system there is a large range of spawn timing among runs. For example, the Early Stuart runs in the Fraser River occur in summer whereas the Cultus Lake run occurs in late fall. These differences in spawn timing result from adaptation to the temperatures of the spawning habitats. Eggs develop quickly in warmer streams and are therefore laid later so that fry emerge at similar times to coincide with optimal feeding conditions when they migrate downstream to their rearing lakes.

The rearing lakes often contain juvenile sockeye from several distinct populations that spawn at different times in different territories, and those mixed populations of sockeye juveniles can also intermingle with juvenile kokanee. Yet, the populations, and the sockeye and kokanee, remain distinct, with high fidelity to spawning sites permitting minimal interbreeding. The juvenile sockeye can spend one to three summers in the lakes, foraging on aquatic invertebrates, particularly waterfleas (*Daphnia*) before migrating downstream to sea. The kokanee, however, remain in lakes. Most sockeye populations have a juvenile stage that rears in lakes, though a few populations inhabit river systems that do not have lakes and some populations migrate to sea in the first year of life.

Sockeye may spend one to three years at sea where they are thought to compete with pink salmon for food resources. At sea, sockeye feed on smaller organisms such as squid, zooplankton (such as krill, amphipods and copepods) and larval/juvenile fishes such as sand lance, herring, pollock and capelin. Individual spawning runs can be composed of individuals of different ages. Larger and older spawners enjoy greater reproductive success on the spawning grounds but risk higher mortality at sea while they feed and grow. Some populations have little variation in freshwater and ocean residency, such as the Adams River run, in which the age at maturing is four years. This run is famous for its four-year cycles of low, intermediate and high abundance that have persisted for decades. The causes for the regular cycles are not known but may involve food web interactions where sockeye from abundant year classes suppress fish from weak year classes, perhaps by boosting predatory fishes in the rivers and lakes.

A school of (mostly) sockeye post-smolts migrates past a salmon farm in Okisollo Channel, BC.

OBSERVING IN THE WILD

The spawning runs of sockeye salmon can be an amazing spectacle of abundance, colour and wildlife. In years of high abundance, the bright colouration of sockeye makes them highly visible, and the wildlife activity of bird and mammalian predators and scavengers creates quite a scene. In some river systems, sockeye provide a major pulse of sustenance for the annual nutrition of bears. Juvenile sockeye, however, are not easily observed in freshwater systems. At sea, juvenile sockeye will co-mingle with pink and chum. Look for surface activity—dimpling and leaping—outside the kelp zone. Sockeye post-smolt leaps are called "sliders," for their lazy jumps that seem to drag over the surface of the water.

CONSERVATION

In their southern range, sockeye are of increasing conservation concern. More than 95% of the former populations in the Columbia River system have been lost to dams. There is one population listed as endangered (Snake River) and one listed as threatened (Lake Ozette) under the Endangered Species Act in the US. The Fraser River stocks of sockeye, historically Canada's largest and most economically valuable salmon fishery, have seen steady declines in recent decades, with losses from climate change, habitat alteration and infectious disease. In British Columbia, the Committee on the Status of Endangered Wildlife in Canada (COSEWIC) lists nine populations as endangered, two as threatened and five as special concern, all of which are from the Fraser River except the Sakinaw Lake population. The status of sockeye improves as you move north, with the Bristol Bay fishery a flagship example of sustainable fisheries management. The species is abundant in the North Pacific Ocean, thriving in the middle to northern ranges but under multiple threats in the southern ranges.

Coho Salmon

(Oncorhynchus kisutch)

Silver Salmon, Hooknose Salmon

Coho salmon are the third largest by body size, behind Chinook and chum, and the second least abundant of the Pacific salmon, not including sea-run trout. The origin of the name "coho" is unknown but is believed to be derived from an Indigenous word for the fish. Their coastal distribution during their marine phase, their willingness to strike at a fly or lure in freshwater and their leaping, dodging and aggressive fight make coho highly prized by sport fishers.

Coho are somewhat different from most other Pacific salmon in that they often do not undertake large oceanic migrations, but instead remain in coastal waters before returning to spawn. They are abundant in small coastal streams and in some large rivers. Juvenile coho typically rear in freshwater streams for one to three years, where they establish territories to feed on larval and adult insects drifting in the current. Survival of juveniles that do not establish territories is poor, which means that rivers tend to produce a similar number of smolts each year, even if the abundance of spawning adults is highly variable. The main predators of coho fry and parr are other trout, whitefish, birds and small mammals. At sea, young coho can feed in groups on smaller fishes but not all coho have the same strategy— smaller coho tend to catch prey themselves whereas larger coho often steal food caught by smaller group members despite being able predators.

Coho typically spend between zero and two winters at sea prior to their spawning migration, which occurs in fall and extends farther upstream than other salmon in small coastal watersheds. Exceptions of long-distance migrants in larger rivers occur, and those fish tend to have a more streamlined body shape than their coastal counterparts. Jacks, the youngest and smallest male spawners, obtain mating opportunities not by competing for mates but by waiting for a female to commence spawning with a larger male and then sneaking in to fertilize some of the eggs. Male spawners develop deep red colouration and deeply hooked upper jaws, which can make them look like Chinook salmon, with which they are often confused. Females and jacks are not as brightly coloured. Adult coho can live up to six years.

Coho are highly sought after by sport fishers in coastal marine waters as well as during their spawning migration in coastal rivers. In rivers coho will strike at flies and lures, even though they are not feeding, perhaps as an act of aggression. They were historically harvested in commercial troll fisheries, but most commercial fisheries have declined or closed over conservation concerns. Their ubiquity in coastal environments make them a primary food fish for many Indigenous Peoples, who harvest them, sometimes using traditional methods such as weirs, nets and gaffs. Coho are a popular food fare with a high fat content supported by a large shrimp component of their diet, but Chinook and sockeye are both fattier.

DISTRIBUTION

In the eastern North Pacific, the spawning distribution of coho extends from Monterey Bay in California to the Aleutian Islands up to Point Hope, Alaska. They are also found in northern Japan and eastern Russia, including the Kamchatka Peninsula and rivers bordering the Sea of Okhotsk. In most regions, coho spawn in smaller coastal rivers and streams, where their freshwater migration is relatively small (usually less than 300 km or less than 200 mi, which is farther than pink and chum salmon but not as far as Chinook and sockeye). However, in other regions they migrate up larger river systems, sometimes great distances—such as more than 2000 km (1200 mi) up the Yukon River and 800 km (500 mi) up the Columbia River—and arrive in small tributaries where they spawn. The marine distribution of coho salmon is coastal, usually within the continental shelf. Coho have been introduced outside their native range to the North American Great Lakes.

LIFE CYCLE

Coho return to their natal rivers in July and August in northern Alaska, September and October in British Columbia and November and December in California. The freshwater upstream migration often begins in response to increased river flow and a high tide. Coho do not have multiple spawning runs per river as do Chinook. Eggs hatch and fry emerge the following spring and begin one to three years of freshwater residency, with more northern populations having longer freshwater rearing periods. Juveniles are territorial and competition for space, which provides drift-feeding opportunities and cover from predation, is intense. Fry and parr experience high predation pressure from larger fishes, birds and small mammals.

Migration to sea begins in spring. As they grow, smolts shift their diet more towards smaller fishes and away from marine invertebrates. Predation by coho post-smolts on smaller pink and chum salmon during their early marine life stage as well as on other small fishes, such as juvenile sand lance, can be high. As coho grow, they increasingly consume more varieties of fish species, particularly herring and sand lance but also rockfish, sardine, anchovy, capelin, pollock and hake. The adult coho has a similar diet to that of Chinook but contains much more invertebrate prey (20% versus less than 1%), including krill, crab larvae, squid, jellyfish and even goose barnacles.

The marine phase of the coho life cycle remains in coastal waters, unlike most other Pacific salmon species, which migrate to offshore waters. Most coho

A lone coho salmon next to Lelu Island in the Skeena River estuary, BC

spend just one winter at sea, except for jacks that return to spawn the same year they migrate to sea, and a small fraction that return after two winters at sea. Most spawners are therefore four (in the south) or five (in the north) years old, depending on latitude, with spawning occurring earlier in the north and later in the south. With their large body size they can deposit several thousand eggs in their redds. Adult coho are an important food resource for marine mammals, such as seals and sea lions and terrestrial mammals, such as bears.

OBSERVING IN THE WILD

Coho fry and parr are easily observed by snorkeling in small coastal rivers and small tributaries of larger rivers, and parr can be caught by angling. Post-smolts can be observed via surface activity, such as dimpling and leaping in coastal marine habitats intermixed with similarly sized pink, chum and sockeye post-smolts near the shore but outside the kelp zone. It is possible to see coho post-smolts foraging on smaller fishes—look for eruptions of surface activity of the smaller fishes as the coho strike from below. Juvenile and adult coho leaps are distinguished by a high athletic arc and clean entry. Adult coho can be caught by trolling in coastal marine waters as well as by fly-fishing and spin casting in rivers. Their relatively large body size within small rivers as well as their remarkable colouration and aggression make adults easily visible in their spawning habitats.

A coho fry near Princess Royal Island, BC

CONSERVATION

Coho share the distinction with Chinook of being among the most imperilled of Pacific salmon, particularly in their southern range in California, Oregon and Washington, mainly because of dams, irrigation, habitat alteration and overfishing. They have been extirpated from large sections of their inland ranges in southern rivers such as the Columbia and Sacramento rivers. In the United States one population is listed as endangered and three are listed as threatened under the Endangered Species Act, all in California, Oregon coastal sections and the lower Columbia River. In Canada, one population is listed threatened by the Committee on the Status of Endangered Wildlife in Canada (COSEWIC)—the Interior Fraser population. In the southern portion of their range, coho abundance has been supported by hatchery production, which in recent decades has accounted for most of the population relative to naturally spawned individuals, creating concerns for the genetic integrity of remaining wild populations in those areas.

Chinook Salmon

(Oncorhynchus tshawytscha)

King Salmon, Spring Salmon, Blackmouth Salmon, Tyee Salmon

Chinook salmon are the least abundant but largest of all Pacific salmon and are named after the Chinook people of the Columbia River. Fish larger than 14 kg (30 lbs) are known as "tyee salmon." *Tyee* means "chief" in the language of the Nuu-chah-nulth First Nation on the west coast of Vancouver Island. The name "spring salmon" refers to the tendency of some Chinook populations to return to rivers in spring, earlier than the summer and fall migrations of other Pacific salmon species. Different populations of Chinook salmon return to rivers at various times throughout the year, and large southern rivers often have multiple seasonal runs. For example, the Fraser River and Columbia River have three discrete runs that peak in spring, summer and fall, and the Sacramento River historically had four seasonal runs. Northern rivers such as the Yukon River tend to have a single early run that peaks in late spring.

Chinook vary in the number of years they rear in freshwater (including zero, like pink and chum) and can spend up to five years at sea. The oldest Chinook mature at eight years of age and tend to be female. The largest Chinook salmon on record was 57 kg (126 lbs), caught in a fish trap near Petersburg, Alaska, in

1949; the largest sport-caught Chinook was 44 kg (97 lbs) caught in the Kenai River, Alaska, in 1985. Historical photographs from the 1800s and early 1900s throughout the species range are replete with examples of such massive fish, however modern examples are increasingly rare. The species has declined in body size as well as age structure, perhaps as an evolutionary consequence of high mortality of the largest fish from size-selective fisheries as well as predation by marine mammals whose populations have increased after their overexploitation in the late 1800s and early 1900s.

Jacks, the smallest spawning Chinook, are males that return to spawn after only a few months at sea. They are much smaller than the older and larger mature females and males, and they have a rather clandestine reproductive strategy: instead of competing with large males for mates, they sneak in and fertilize some of the eggs when a female is mating with another fully grown male. Because they do not compete for mates, jacks devote much more of their body mass to sperm production in comparison to large males, who instead risk years at sea to invest in a large body size that gives them an advantage when competing for mating opportunities. Jacks also occur for coho salmon, less commonly for sockeye, rarely for chum, and not at all for pinks.

Historical records suggest that Chinook runs at the southern and northern edges of their range were similar or even more abundant than runs in the centre of their range. Those peripheral, but abundant, stocks were and are of central importance to food security and culture of Indigenous Peoples of the Pacific Northwest. The large body size of Chinook salmon makes them a major focus of recreational fisheries. “Ivory king” is the name sometimes given to certain individual Chinook salmon that have white fillets owing to a genetic inability to take up the pink carotenoids from shrimp and krill and are prized in restaurants and seafood markets.

DISTRIBUTION

The historical range of Chinook salmon extended from California to Alaska, with highly abundant runs in the Sacramento–San Joaquin River System in the south and the Nushagak and Yukon rivers in the north. Runs also occur in parts of eastern Russia, particularly the Kamchatka Peninsula. Multiple factors, including irrigation, dams, habitat alteration and overexploitation, have caused the range of Chinook salmon to contract in the south. Chinook often migrate thousands of kilometres up major river systems and can spawn in the mainstem or headwaters of large rivers as well as smaller tributaries and coastal rivers. Chinook have been introduced and become established beyond their native range; their worldwide distribution now includes the Great Lakes, Patagonia and New Zealand.

LIFE CYCLE

The life history of Chinook salmon is extremely variable, and consequently there are exceptions to any general pattern. Spawn timing is typically earlier in the year in the north, occurring as early as May or June for northern stocks in Alaska, Yukon and British Columbia, and then ranging from November to January in Oregon and California. However, many rivers, such as the Sacramento–San Joaquin, Columbia and Fraser rivers, have multiple discrete spawning runs seasonally over the year. To reach their spawning grounds, Chinook can migrate upstream more than 3000 km (nearly 2000 mi), travelling 45 km (28 mi) per day and surviving primarily on their energy reserves.

With their large body size, Chinook can deposit more than 10,000 eggs in redds that cover several square metres and are as large as a Volkswagen Beetle. After emerging from their redds, some juveniles rear in rivers and streams for a year or more before migrating to sea (especially in the north), whereas others will spend only a month or two (especially in the south). The former are called stream-type Chinook, the latter (which migrate to sea as fry, like pink and

Young Chinook salmon captured during their coastal migration in Okisollo Channel, BC

chum) are called ocean-type Chinook and often co-occur with their stream-type counterparts in the middle of the species' range.

In freshwater, juvenile Chinook eat a lot of insects; once in the ocean they switch to marine invertebrates and smaller fishes, such as herring, sand lance and pilchard as the Chinook increase in body size. The marine residency time for Chinook ranges from a few months for jacks to five years. The marine distribution of Chinook salmon is broad, sparse and variable. Ocean-type Chinook tend to remain in coastal marine waters whereas stream-type Chinook tend to move to offshore waters of the Gulf of Alaska and Bering Sea. The coastal distribution of Chinook is an important food source for coastal marine mammals, particularly resident-type orcas.

OBSERVING IN THE WILD

Chinook fry and parr can be observed in rivers and streams feeding on insect larvae, and adults in benthic habitats and drifting in the current. At sea, juvenile and adult Chinook have a broad and sparse distribution, making them rare to see other than by anglers, but they are conspicuous in spawning habitats. Spawners and their VW Beetle-sized redds can be observed in mainstem rivers under the right conditions of water clarity and light.

A scarred Chinook spawner in the Quinsam River, BC

CONSERVATION

Chinook salmon are among the most imperilled of the Pacific salmon. They have been extirpated from numerous sections of the southern portion of their range. From California to Washington, there are two stocks listed as endangered and seven stocks listed as threatened under the Endangered Species Act. In British Columbia, there are two stocks listed as special concern, seven listed as threatened and 13 listed as endangered by the Committee on the Status of Endangered Wildlife in Canada (COSEWIC). There are no Chinook stocks listed under the Endangered Species Act in Alaska. The primary threats to Chinook salmon have been dams, irrigation, habitat alteration and overexploitation. For some at-risk stocks, artificial propagation through hatcheries has helped to support abundance but has contributed to genetic deterioration. Conservation of Chinook salmon is of importance for the conservation of imperilled resident orcas who rely on Chinook salmon as a food resource.

Steelhead Trout

(Oncorhynchus mykiss)

Sea-run Rainbow Trout, Hardheads

Steelhead are the large sea-going form of rainbow trout. The origin of the name "steelhead" is not known but is likely a reference to the bright steel-blue colouration this fish displays while at sea as well as the fish's blunt head. Steelhead are a major conservation concern in their southern native range; however, rainbow trout are globally the most successfully introduced salmonid, invading many continents beyond their native range. There are particularly unique forms of rainbow trout in the southern parts of their range, such as the redband trout of the Columbia and Sacramento rivers, golden trout of the Kern River in California and the Rio Mayo and Guzman trout of Mexico, all of which have spectacular colouration and spotting patterns. Scientists disagree whether the golden, Rio Mayo and Guzman trout are unique species or locally adapted populations.

There are many groupings (arguably subspecies) of rainbow trout in their native range, but steelhead only occur in some of them. Most steelhead spawn in coastal rivers and streams from southern California to Bristol Bay. The interior steelhead, including the notable populations of the Columbia and Fraser rivers, migrate farther inland and take on a much deeper red colouration along their lateral line and ventral surface. Lake-adapted steelhead-like forms also occur in the native and introduced ranges, such as the redband trout of Upper Klamath Lake Basin (*O. m. newberrii*) where they migrate to and utilize large lakes like steelhead use the ocean. Unlike anadromous sockeye and freshwater-resident kokanee, which do not interbreed, the anadromous and freshwater forms of rainbow trout do interbreed, albeit infrequently. The offspring of freshwater-resident mothers can become anadromous and vice versa. Interbreeding can occur via "sneaker males," which in this case are small freshwater-resident trout (not jacks) that surreptitiously and opportunistically fertilize some eggs when a steelhead pair is spawning. Interbreeding can occur when juvenile steelhead males become sexually mature before migrating to sea and spawn with a similarly sized female trout.

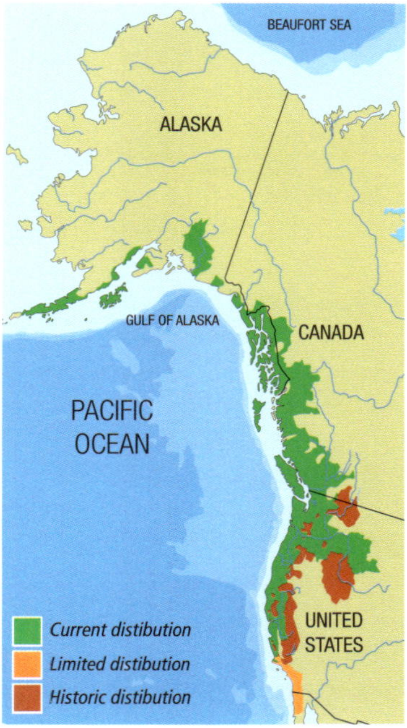

Steelhead, like coastal cutthroat, differ from other Pacific salmon because they spawn in spring and can spawn multiple times before they die, returning to the ocean to feed and grow between spawning. In their native range, repeat spawning is relatively uncommon, occurring around 10% of the time. In their introduced range, such as Lake Superior, 70% repeat spawn, perhaps because there are fewer predators. At sea, steelhead can roam for thousands of kilometres, but not always. In Kamchatka, Russia, there is an estuarine form that stays near its natal river, an anadromous form that feeds far offshore for several years and an anadromous form that returns immature in its first year at sea to overwinter before going back to sea again for several more years. Amazingly, all these life histories can occur among siblings.

Deliberate introductions where they were historically favoured by fishery managers for their hardiness and sport-fishing value as well as unintended introductions from aquaculture facilities have resulted in rainbow trout having one of the broadest non-native ranges of all freshwater fishes. In their non-native ranges, the steelhead and stream-resident forms have continued to coexist, including non-anadromous but steelhead-like forms that migrate from spawning rivers to the Great Lakes for feeding and growth rather than migrating to sea. However, rainbow trout have created ecological havoc in their introduced ranges, outcompeting and predating other native fishes. Commercial fisheries don't exist for steelhead, though they can be caught as by-catch in fisheries targeting other species of salmon. Aquaculture-produced rainbow trout are abundant and pond-produced rainbow trout are considered a more sustainably produced fish than steelhead and other salmon species grown in ocean net-pen farms.

DISTRIBUTION

Steelhead primarily spawn in coastal rivers and streams ranging from southern California to southern Alaska, and typically do not migrate far inland except for redband steelhead that migrate up the Columbia and Fraser rivers. Their ocean migration is highly variable, ranging from a marine phase that remains in coastal waters near their natal rivers to large offshore ocean migrations. The freshwater resident forms have populations as far south as northern Mexico and many have unique characteristics, such as the colourful golden trout of the Kern River, the spectacularly striped and spotted Mexican populations or the exceptionally large Kamloops trout of the large lakes of interior British Columbia. Rainbow trout, and in some cases steelhead, have succeeded in invading freshwater ecosystems beyond their native range throughout North America and parts of South America, Europe, Asia, Russia, Australia, New Zealand and Africa.

LIFE CYCLE

Steelhead spawn in spring, but there are distinct summer and winter runs that migrate from the ocean into coastal rivers. The winter run enters rivers between November and May; the fish are sexually mature and migrate relatively quickly to the spawning grounds. The summer run enters rivers as early as April and continues until the end of November. In contrast to the winter run, fish in the summer run are sexually immature or have not begun sexual development when they migrate into the rivers; they instead spend several months up to a year in freshwater, feeding and maturing sexually prior to spawning.

At spawning, steelhead, like other large salmon, form male and female spawning pairs but can interbreed with local resident rainbow trout when small male residents opportunistically fertilize some eggs from large female steelhead

Steelhead parr in the Heber River, BC, close to its confluence with Gold River

that are spawning. Offspring of steelhead can take on the resident trout or the anadromous steelhead lifestyle, and siblings that become steelhead can vary in the ocean migration patterns they undertake. Juvenile steelhead can spend one to several years in freshwater before migrating to sea, and juvenile males can become sexually mature and interbreed with female resident trout before commencing their ocean migration.

Juvenile steelhead feed on zooplankton and aquatic invertebrates, such as mayflies, caddisflies and black flies; after migrating to sea, steelhead feed on a broader range of species as they grow, including zooplankton, invertebrates and smaller fishes. They typically spend two to three years at sea feeding and growing. However, a small proportion returns immature the next fall to overwinter and then migrates back to sea the next spring, without sexually maturing and spawning. Steelhead can spawn multiple times; those that survive spawning and are in the process of returning to the ocean are called "kelts." Repeat-spawning steelhead tend to spend one or two years feeding in the ocean before returning to spawn, but some do return again the following year. Females survive spawning better than males, and most fish over 70 cm (28 in) are repeat spawners.

OBSERVING IN THE WILD

Adult steelhead are probably most observed by recreational fishers and are particularly sought after and prized by anglers fly-fishing in coastal rivers and streams. Juvenile steelhead and rainbow trout are abundant and can be easily observed in streams, especially by snorkeling, as well as from the riverbanks and by angling. It is not possible to distinguish between juvenile steelhead and rainbow trout, not even for a geneticist. Snorkeling for steelhead and rainbow trout in streams is especially appealing because you can easily be non-disruptive and observe in detail their natural behaviours, such as feeding on drifting invertebrates and fighting over foraging territories.

CONSERVATION

The Columbia River historically had the largest run of steelhead in North America, with nearly five million pounds harvested commercially in 1892. However, dams and now climate change and water diversion have decimated this run to a point where it is primarily supported by hatchery propagation. Half of the remaining stocks in the Columbia River are at a high risk of extinction and 14 stocks are extinct. South of the Columbia River, the conversation status of steelhead worsens, whereas in southern Alaska, the stocks remain healthy. In the United States there is currently one stock listed as endangered and ten as threatened under the Endangered Species Act, all in Washington State and southward. Two steelhead populations are listed as endangered by the Committee on the Status of Endangered Wildlife in Canada (COSEWIC)—the Chilcotin and Thompson River populations.

Coastal Cutthroat Trout

(Oncorhynchus clarkii clarkii)

Sea-run Coastal Cutthroat Trout, Blueback Trout, Harvest Trout, Yellowbellies

Coastal cutthroat is a subspecies of cutthroat trout, which exhibits the greatest diversity of life histories and migration patterns of Pacific salmon and trout. Cutthroat trout are named for the red slashes under their jaws (although sometimes these are orange or yellow) and have a variety of spectacular colourations and spotting patterns that vary among subspecies and life-history stages. Compared to the interior subspecies, coastal cutthroat are more heavily spotted throughout their body, can be silvery like salmon, and have more subdued colouration patterns. The scientific name for cutthroat trout, *Oncorhynchus clarkii*, comes from the Lewis and Clark expedition, which first encountered the fish near Great Falls on the Missouri River in Montana in 1805. Cutthroat trout were initially considered a subspecies of rainbow trout, and both were grouped taxonomically in the genus *Salmo*, which includes Atlantic salmon. However, they are now recognized as two species of the Pacific salmon and trout genus *Oncorhynchus.*

Cutthroat trout are represented by four evolutionary groupings (coastal, westslope, Yellowstone and Lahontan), which may encompass roughly 14 subspecies but the exact taxonomic distinctions of cutthroat trout are

a contested subject. Most of these potential subspecies are localized to specific river basins such as the Lahontan group of five subspecies and the Yellowstone group of seven subspecies, which extend into the interior of North America via the Columbia and Colorado river basins and east of the Continental Divide into the South Saskatchewan, Missouri and Arkansas rivers and Rio Grand. While cutthroat trout may best be known as a freshwater trout in mountain streams that is highly prized by fly fishers, the coastal cutthroat trout is the most abundant subspecies and inhabits coastal systems where its life history includes anadromous and lake-dwelling forms that take on the large body sizes and silver colouration of salmon. Coastal cutthroat trout is the only cutthroat subspecies that has an anadromous, or "sea-run" form, which is the form we focus on in the identification sections of this book.

The life history types include stream residents, lake residents and migratory forms that travel between the headwater streams where they spawn and the large rivers, lakes, estuaries and marine waters where they spend the rest of their lives. Of the migratory forms, only those that travel to estuaries or the ocean are called anadromous, and an individual fish may express multiple migration strategies prior to spawning. Each of these life-history types can coexist within a population, and spawning can occur between all of them; for instance, it is common to observe small male freshwater residents trying to sneakily fertilize the eggs of an actively spawning anadromous female paired with a larger male. Of the easily delineated life-history types, the stream resident is smallest, the lake resident is largest and the anadromous form is intermediate in size Anadromous coastal cutthroat do not undertake large ocean migrations, but rather remain in estuaries or near-shore marine habitats during spring and summer, and typically remain within near from their home river. Some migrate back to their natal river to overwinter but without sexually maturing and spawning, whereas others may overwinter in estuarine or ocean habitats. Like steelhead, coastal cutthroat often survive spawning and can spawn as many as five times in their lifetime. Coastal cutthroat that have spawned and are in the process of returning to the ocean are called "kelts."

Coastal cutthroat trout are smaller and less abundant than other Pacific salmon, but their year-round availability in coastal rivers and streams as well as near-shore marine habitats have made them important to Indigenous fisheries. Cutthroat trout are not fished commercially, but their aggressive feeding and fight make them enormously popular amongst anglers, especially fly fishers. Subspecies of cutthroat trout in their interior range are propagated by hatcheries for restoration purposes and to support recreational fisheries.

DISTRIBUTION

Coastal cutthroat have the broadest geographic distribution among subspecies of cutthroat trout, ranging from Humboldt Bay in northern California to Prince William Sound and the Kenai Peninsula in Alaska. Other subspecies of cutthroat trout have relatively small ranges restricted to river basins but occur from New Mexico to southern Alberta east of the Continental Divide. In contrast to the other Pacific salmon and trout, coastal cutthroat occupy almost all of their known historical range but there is not enough information to characterize regions with "limited" coastal cutthroat presence. Anadromous coastal cutthroat are rarer than the other life-history types in California and near highly populated coastal cities, mainly because of poor quality estuary habitat. In marine waters, coastal cutthroat often remain near their natal river typically ranging less than 80 km (50 mi) into the ocean from their home stream. Although they have been observed farther from shore in the plume of large rivers, they do not occur in open ocean habitats and have rarely been observed farther than 150 km (approximately 100 mi) from the shoreline. Many anadromous coastal cutthroat remain in the estuary of their natal river, not migrating farther into the ocean. Spawning occurs in small tributaries and headwaters of coastal streams.

Coastal cutthroat are furtive fish that like woody debris (Princess Royal Island, BC).

LIFE CYCLE

Coastal cutthroat spawn from winter to early spring, with spawning occurring later in more northern populations. Anadromous coastal cutthroat typically begin congregating in estuaries in late summer, where they remain for several months feeding on smaller fishes and invertebrates before migrating upstream. The timing of upstream migration is consistent within populations but is highly variable among rivers (ranging from July to February) and is later in the north, which is likely related to the distance to spawning areas or other environmental factors such as temperature. Some rivers have distinct spawning migrations, such as fall and winter runs in the Stillaguamish River in Washington. Upstream migrants are a combination of repeat spawners, first-time spawners and sexually immature fish. Typically, older fish, which carry more eggs as well as larger eggs, move into rivers earlier than younger fish. Hybridization with rainbow trout can occur.

Repeat spawning is common, with up to 40% of spawners surviving to return to sea in some populations. The record is five spawnings by a single 10-year-old fish. The proportion of freshwater migrants that are sexually immature and migrate upstream to overwinter rather than spawn varies among populations but is more common in northern rivers where those fish usually spawn on their second upstream migration. Female anadromous coastal cutthroat usually become sexually mature by four years of age and have a lifespan of seven to nine years. However, the stream-resident form, particularly in small streams above waterfalls, rarely lives more than three years.

Why the anadromous form is smaller than the lake-dwelling form is a mystery but may be a consequence of the shallow and low-flow habitats used by anadromous coastal cutthroat trout for spawning and rearing to minimize interactions with other salmonids. Anadromous coastal cutthroat are smaller than steelhead, whereas lake-resident coastal cutthroat, where they coexist with lake-resident rainbow trout, are larger. These lake-adapted cutthroat prey mainly on fishes, and the lake-adapted rainbow trout, which would otherwise also feed on smaller fishes, instead restrict their diet mainly to invertebrates where they share the lake with cutthroat trout.

Coastal cutthroat hatch and emerge from redds in the spring, several weeks after spawning. The fry move into low-velocity waters of side channels, where they rear for their first summer before moving into low-velocity mainstem sections of small streams and feed primarily on invertebrates. The age of smoltification and downstream migration is highly variable within and among populations, ranging from one to six years, but is usually between two and four years. Once anadromous coastal cutthroat reach estuaries and the ocean, they grow rapidly, and their diet expands to include various crustaceans and smaller fishes, such as stickleback, sand lance and herring.

OBSERVING IN THE WILD

Coastal cutthroat are most observed by recreational fishers and are highly prized by anglers fly-fishing or spin-casting in coastal streams, rivers, lakes, estuaries and nearshore habitats throughout the year and especially during summer and fall in estuaries as the fish congregate for their upstream migration. Juvenile and adult coastal cutthroat can be observed in rivers and streams by snorkelers. To find juveniles, look in small, low-velocity streams. Find adult migrants by snorkeling in mainstem rivers or simply by walking along small tributaries and headwater streams during the spawning season.

CONSERVATION

Their dependence on small tributaries and headwater streams makes coastal cutthroat highly sensitive to many human disturbances including dams, logging, urbanization and agriculture. They are considered a sentinel species of aquatic system health, often being the first salmonid to be lost in response to habitat degradation and the last to recover following habitat restoration. Their natural ability to hybridize with rainbow trout has resulted in domesticated rainbow trout genes arising from artificial propagation in hatcheries invading some coastal cutthroat populations where widespread stocking of rainbow trout occurs. In addition, stocking has resulted in hatchery-produced cutthroat interbreeding with wild coastal cutthroat with negative consequences for the growth and survival of the subsequent generations.

A steelhead smolt in the Muchalat River, BC

Both anadromous and non-anadromous populations of coastal cutthroat have steadily declined in California, Oregon, Washington and British Columbia. In 1996, the coastal cutthroat of the North Umpqua River, Oregon, was listed as endangered but then delisted in 1999 when they were regrouped with the broader Oregon Coast Evolutionary Significant Unit, which was not considered at risk of extinction despite population declines, habitat degradation and interbreeding risk. Coastal cutthroat in British Columbia are not listed by the Committee on the Status of Endangered Wildlife in Canada (COSEWIC) but are listed as a species of special concern by the provincial government.

IDENTIFICATION BY LIFE STAGE

FRY AND PARR

Young salmon spend most their time in freshwater as fry and later, for some species, as parr. Some species can stay in freshwater for more than one year, at which point the fish are called parr rather than fry. Freshwater residence of juveniles is a crucial part of the salmon life cycle for many reasons, a main one being that this is when salmon imprint on their natal streams so they can return to spawn later in life. Salmon fry and parr must learn quickly how to avoid predators while still obtaining food. Like many fishes, salmon are born into the world without any form of parental care.

When a salmon hatches from its egg, usually in the early spring, it is called an alevin and has a yolk sac attached to its belly that it will depend on for nourishment over the coming weeks or months. Alevin remain hidden in the gravel close to where they hatch, gradually absorbing their yolk sac until only a slit in the belly remains—a process called "buttoning up" or "zipping up." Alevin then emerge from the gravel and swim to the surface to gulp air to fill their swim bladder, giving them buoyancy and allowing them to swim above the gravel with far less difficulty. Gravel emergence marks a salmon's introduction into the world beyond its redd (nest) and its transition from alevin to fry.

Coho, Chinook, coastal cutthroat and steelhead fry often travel downstream after gravel emergence and settle in calm sections of water, such as eddies or pools, where they will remain until they are ready to migrate to sea or are displaced by water flow or lack of food. In contrast, sockeye fry emerge from gravel and soon migrate into their nursery lake, where they will spend the next year of their lives; this mini migration can be downstream or upstream, depending on the population, and typically occurs at night. Some sockeye, coho and Chinook will migrate to saltwater soon after gravel emergence, with some populations heavily favouring this life-history strategy, but this is less common than spending a year in freshwater. Coho and Chinook that migrate to the ocean as fry often make use of estuaries as nursery habitat for weeks or months.

Pink and chum are unique in that they *always* immediately migrate to the ocean after emerging from the gravel. With less time to prepare for ocean life, they are barely ready for saltwater when they enter it and are often still "unzipped" with their belly slits or even some of their yolk sacs still visible. This transition is so quick that they look much the same in the ocean as they did in the river just days before. The line between fry, smolt and post-smolt is therefore blurred for pink and chum. The illustrations for pink and chum fry are representative of any unscaled individuals after yolk absorption, regardless of water salinity.

Fry are roughly 2–6 cm (1–2.5 in) in length and parr are generally around 6–15 cm (2.5–6 in), depending on the species, but coastal cutthroat and steelhead parr can be even larger. Like smolts, parr have scales and a more streamlined snout and body, but they do not silver up and undergo the process of smoltification until they are ready to migrate to sea. Some sockeye, coho and Chinook will remain in freshwater beyond that first year, but most do not; and when they do stay behind as parr, they look like smolts, just less silvered. Coastal cutthroat and steelhead, on the other hand, *usually* stay in rivers and look quite different as parr compared to smolts. Consequently, we have included parr illustrations for coastal cutthroat and steelhead in this section, but not for the other species.

Except for pink and chum, salmon fry are benthic creatures, meaning they spend most of their time close to the bottom of their stream or lake. Their colouration and parr marks (the dark vertical markings on the flanks of fry and parr) help them camouflage with the surrounding rocks, debris and shadows. Fry are more solitary and aggressive than smolts, eschewing schools for a more independent and inconspicuous existence. This is especially true for coho and steelhead. Individual fry of these species usually occupy and defend small feeding territories by chasing and nipping other fry that come too close. Salmon fry mostly eat insects (larvae, nymphs and adults) and zooplankton. Small and abundant, young salmon in streams are themselves a choice meal for many birds and larger fishes, including older salmon.

If you have ever seen a small fish dart between hiding places in a calm, shallow section of creek in the Pacific Northwest, there is a good chance that it was a salmon (or trout) fry or parr. Although they are small and secretive, fry and parr are 100 to 1000 times more abundant than the spawners that will eventually return, so they are certainly there to be seen. Patient observers on the edges of salmon-spawning creeks can be rewarded with year-round sightings of fry and parr searching for food or zipping among rocks. Even if you're unable to get a perfect view of the fish, adipose fins and parr marks are dead giveaways that you're looking at a salmonid, and some species have clear distinguishing traits as well. Fry and parr are typically caught via minnow traps or small seine nets, primarily for research purposes since they have no food value and trapping them for bait is illegal.

Salmon fry and parr have exquisite colours and patterning. From throat slashes to bright orange fins, the prominent markings of fry and parr are at odds with the homogeneity of salmon in the ocean. These markings will eventually fade or disappear as fry and parr silver up into smolts and post-smolts, but for now they allow for relatively easy distinction among species, except for coastal cutthroat and steelhead fry.

SUMMARY TABLE OF FRY AND PARR KEY CHARACTERISTICS

Feature	Pink	Chum	Sockeye
Habitat	freshwater or marine	freshwater or marine	freshwater
Parr marks	none	shorter than vertical diameter of eye; mostly above lateral line; height regular	most are shorter than vertical diameter of eye; slightly more above lateral line than below (but straddles more than chum); height staggered and irregular
Dorsal spotting	none	smaller fish (<5 cm or 2 in) have regular row of spots between top of parr marks and dorsal ridge; larger fish (>5 cm or 2 in) have irregular spots and blotches	distinct, irregular, large, well-spaced spots that fade as fish grows
Dorsal fin	mostly colourless	mostly colourless, but gains a dusky spot on tip at larger sizes (>5 cm or 2 in)	mostly colourless, but occasionally with a yellow or green hue and/or a half-height black leading edge
Adipose fin	colourless	colourless	colourless
Anal fin	colourless	colourless	colourless
Other key features	none	dorsal spotting variable, but present; in freshwater, flank below lateral line sometimes has greenish sheen	none
Most similar species	unique	Chinook and sockeye	chum and Chinook

Coho	Chinook	Steelhead	Coastal Cutthroat
fresh or brackish water	fresh or brackish water	freshwater	freshwater
longer than vertical diameter of eye; straddle lateral line; height regular	longer than vertical diameter of eye; straddle lateral line; height regular	longer than vertical diameter of eye (in fry, but not necessarily in parr); straddle lateral line; height regular	longer than vertical diameter of eye (in fry, but not necessarily in parr); straddle lateral line; height regular
many muted and irregular spots around dorsal ridge, with larger spots nearer and between the parr marks	prominent, but highly variable in size, shape and placement	large dark spots between top of parr marks in fry, which transition to smaller leopard-like spots in parr, mostly away from ridge; dorsal parr marks may be present along dorsal ridge ahead of dorsal fin, numbering roughly 5 (but up to 10) in fry, but faded and fewer in parr	large dark spots between top of parr marks in fry, which transition to smaller leopard-like spots in parr, mostly away from ridge; rarely any dorsal parr marks along dorsal ridge ahead of dorsal fin; when present, fewer than 5
leading edge long and white, with black just behind	overall colour variable; smaller fish (<4 cm or 1.5 in) have fine black speckles at base of leading edge, which gradually form black leading edge as fish grows	black leading edge and white tip in fry; spotted with white tip in parr	black leading edge and white tip in fry; spotted with white tip in parr
orange or brown, often with black trailing edge	usually has clear, unpigmented centre, with black speckling or partial line along edge	opaque, usually with 1–2 partial black lines along edge	opaque, usually with 1–2 partial black lines along edge
sickle-shaped and orange-brown, with a long, white leading edge and black bar just behind	colourless in smaller fish (<4 cm or 1.5 in); variable colour in larger fish (>4 cm or 1.5 in)	partially orange (with white tip in parr)	partially orange (with white tip in parr)
pectoral, pelvic and caudal fins orange	when out of water, flank sometimes has purple sheen	snout from side rounded in parr; maxillary does not reach rear margin of eye in parr (unlike coastal cutthroat); pectoral, pelvic and caudal fins partially orange	snout from side rather pointy in parr; maxillary usually extends to or past rear margin of eye in parr (unlike steelhead); pectoral, pelvic and caudal fins partially orange; yellow, orange or red slashes under jaw in parr, but not fry
unique	chum and Chinook	coastal cutthroat	steelhead

Pink Fry

A spotless and fragile fish

KEY CHARACTERISTICS

Parr marks and dorsal spotting absent. Flank silver. Dorsal surface colour dependent on the environment (bluer in the ocean, browner in turbid rivers) but generally greener than other species in the same habitat and often gains dark blue patches when fish is stressed from handling. Anal and adipose fins colourless and other fins mostly colourless. Smaller individuals (<3.5 cm or 1.5 in) often "unzipped." *Typical dead fish diagnostics: 27–33 long, slender rakers on first gill arch, 130–195 long, slender pyloric caeca, 14–16 anal fin rays, 11–14 branchiostegal rays.*

BEHAVIOUR AND HABITAT

Migrates to the ocean immediately after emerging from the gravel, so mostly seen in the coastal marine environment. Occupies nearshore habitat, often in just 2–5 m (6.5–16.5 ft) of water, swimming just below or at the surface. Uses kelp as protection when possible. In freshwater, usually found in lower reaches of streams and main stems of larger rivers. Forms schools with co-migrating chum, sometimes in schools of thousands. Easily visible from above in calm water, especially over light-coloured substrates. Makes a characteristic dimpling pattern on the surface when swimming, like droplets hitting the water at low speed.

SIMILAR SPECIES

A pink fry is unique—it is the only salmon fry lacking parr marks and dorsal markings.

Chum Fry

A petite fish with exotic dorsal patterning and seemingly half-finished parr marks

KEY CHARACTERISTICS

Parr marks shorter than vertical diameter of eye, mostly above lateral line and of regular height. Flank below lateral line sometimes has greenish sheen in fresh water. Dorsal surface spotting present, but variable and dependent on size; smaller individuals (<5 cm or 2 in) usually have a regular row of spots between top of parr marks and dorsal ridge, but these give way to irregular spots and blotches in larger individuals. Dorsal ridge stripe evident, but sometimes faint. Dorsal fin mostly colourless; tip of fin gains a dusky spot in larger individuals. Anal and adipose fins colourless. Other fins mostly colourless. Very small individuals (<3.5 cm or 1.5 in) often "unzipped." *Dead fish: 20–26 short, blunt rakers on first gill arch, 160–185 pyloric caeca, 13–14 anal fin rays, 13–14 branchiostegal rays.*

BEHAVIOUR AND HABITAT

Immediately after developing into fry, chum travel to the ocean. Usually found close to the surface in shallow, nearshore marine habitat. Sometimes milling about near kelp fronds. In freshwater, usually found in lower reaches of streams and main stems of larger rivers. Migrate in schools of varying sizes alongside co-migrating pink fry. Surface activity mostly limited to dimpling.

SIMILAR SPECIES

Can be mistaken for a Chinook in freshwater, but a Chinook fry has parr marks that extend farther below the lateral line than on a chum fry. The adipose fin lacks speckling, unlike Chinook, and does not have the purple iridescence sometimes seen on the flank of an out-of-water Chinook fry. Sockeye fry can look similar, but mainly occupy lakes where chum fry are absent. A sockeye fry has irregular, staggered parr marks and lacks the faint green sheen below the lateral line that is sometimes present on the flanks of chum fry in freshwater.

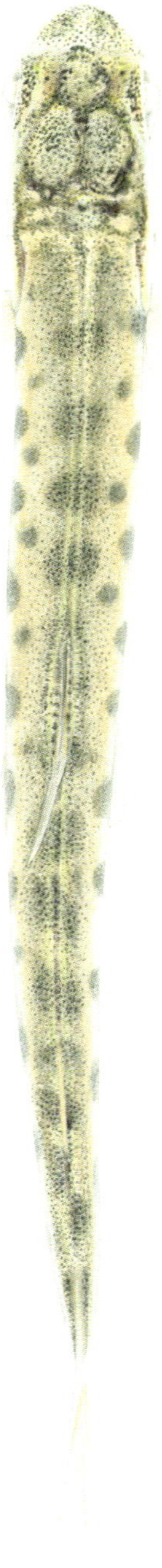

Sockeye Fry

A delicate lake-dweller with haphazard parr marks that look as if they have been hastily painted

KEY CHARACTERISTICS

Parr marks staggered, irregular in height and mostly shorter than vertical diameter of eye. Parr marks that straddle lateral line do so unequally, extending roughly only one-third below line. Dorsal surface usually has distinct, irregular, well-spaced and relatively large spots that fade as fish grows. Dorsal fin mostly colourless, but occasionally has a yellow or green hue and/or a half-height black leading edge. Anal and adipose fins colourless. Other fins mostly colourless, but caudal fin occasionally has a yellow or green hue. *Typical dead fish diagnostics: 32–37 long, slender, close-set rakers on first gill arch, 65–95 pyloric caeca, 14–16 anal fin rays, 13–15 branchiostegal rays.*

BEHAVIOUR AND HABITAT

Primarily found in lakes. After emerging from the gravel, travels slightly downstream to calm areas by stream edges and congregates with other sockeye fry, sometimes in the hundreds of thousands. After a few hours to a few weeks, migrates downstream or upstream to rearing lake, travelling along stream margins. Once in lakes, tends to occupy nearshore habitat for a time but moves deeper as it grows larger to avoid predators, eventually becoming entirely pelagic to feed on zooplankton in deep water. Forms foraging schools with other sockeye fry in lakes.

SIMILAR SPECIES

Chinook and chum can be similar, but the irregular, staggered parr marks of sockeye fry are distinctive. Parr marks extend slightly farther below the lateral line than on chum, but not as far as on Chinook. A sockeye fry lacks the green sheen of a freshwater chum fry and the purple sheen of an out-of-water Chinook fry that is sometimes present on the flank.

Coho Fry

A showy and distinctive fish with eye-catching anal and dorsal fins

KEY CHARACTERISTICS

Parr marks regular in height, straddling lateral line and mostly longer than vertical diameter of eye. Dorsal surface covered in muted and irregular spots around ridge and larger spots nearer and between parr marks. Dorsal spots become smaller and more numerous as fish grows. Dark dorsal ridge stripe sometimes present. Dorsal fin has long, white leading edge, with a thick black bar just behind. Anal fin sickle-shaped and orange-brown with a long, white leading edge and black bar just behind. Adipose fin orange, often with a black trailing edge. Other fins orange. *Typical dead fish diagnostics: 19–27 short rakers on first gill arch, 50–85 pyloric caeca, 13–14 anal fin rays, 13–14 branchiostegal rays.*

BEHAVIOUR AND HABITAT

Occupies all available reaches of streams, including estuaries, but prefers low-velocity areas like stream margins, debris piles and small pools. Mostly solitary (rarely schools) and inactive at night. Feeds higher in the water column and eats more terrestrial and drifting prey than do fry of other species. Occupies and defends small feeding territories through aggressive techniques including posturing, nipping and chasing. Successful individuals will obtain the best feeding opportunities for insects, zooplankton and other benthic or drifting prey. This territoriality results in a somewhat even distribution of fry throughout calm sections of streams and can lead to some fry not being able to obtain space and food. Displaced individuals are often forced to migrate downstream early, either to another section of the stream or all the way to the estuary.

SIMILAR SPECIES

The distinctive sickle-shaped anal fin with a white leading edge is unique to coho fry.

Chinook Fry

A nondescript and unassuming fish with parr marks that firmly straddle the lateral line

KEY CHARACTERISTICS

Parr marks regular in height, straddling lateral line and mostly longer than vertical diameter of eye. Flank sometimes has purple sheen when fish is out of water but alive. Dorsal spotting prominent but highly variable in size, shape and placement. Dorsal ridge stripe dark and usually prominent. Dorsal fin usually has fine black speckles at base of leading edge in smaller fish (<4 cm or 1.5 in), which gradually form a black leading edge as individual gets larger. Anal fin colourless in smaller fish (<4 cm or 1.5 in) and gains pigment in larger fish (>4 cm or 1.5 in). Adipose fin has clear, unpigmented centre and black speckling or a partial line along part of edge. Dorsal, pectoral, pelvic and caudal fins variable in colour, with stream-type individuals having brighter fins than ocean-types. Smells particularly fishy. *Typical dead fish diagnostics: 20–25 short rakers on first gill arch, 140–185 pyloric caeca, 15–19 anal fin rays, 16–18 branchiostegal rays.*

BEHAVIOUR AND HABITAT

Mainly found in large river systems, usually in low-velocity sections of the main channel or tributaries. Like coho, has individualistic feeding behaviour, but much less aggressive. Tends to avoid the slack-water pools preferred by coho, instead occupying fast riffles and deeper water. Some individuals and populations migrate to sea as fry, smoltifying quickly and spending the following weeks to months growing in tidal marshes of estuaries. These ocean-type Chinook are less aggressive than their stream-type counterparts and tend to orient themselves with the current rather than against.

SIMILAR SPECIES

Sometimes confused with chum or sockeye. Parr marks are more regular on Chinook fry than sockeye and extend slightly further below the lateral line than on sockeye and much further than on chum. Unlike chum and sockeye, a Chinook fry has black speckling or dark outline on the trailing edge of the adipose fin and sometimes has a purple sheen on the flank when out of water. Sockeye fry are primarily found in lakes, which Chinook fry do not inhabit.

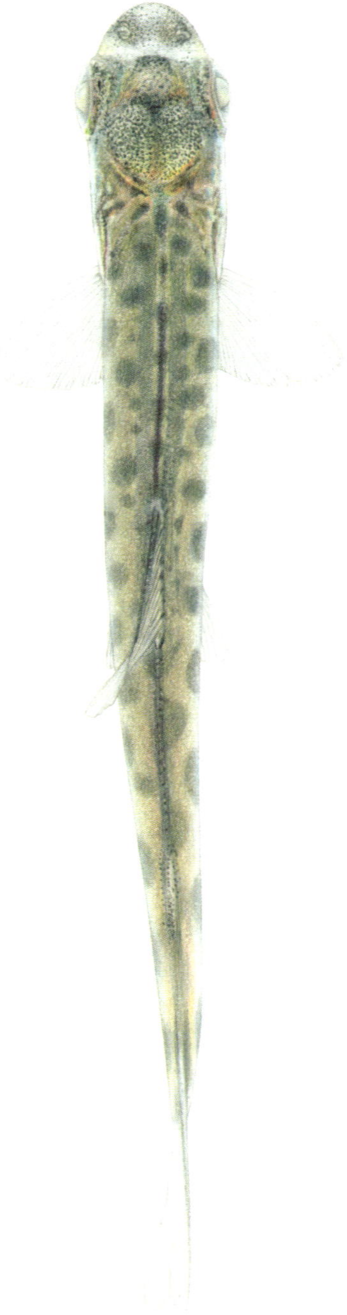

Steelhead Fry

A bit of a bully, zooming along the streambed picking fights with any other fry that gets too close

KEY CHARACTERISTICS

Maxillary does not reach rear margin of eye. Parr marks regular in height, straddling lateral line and mostly longer than vertical diameter of eye. Rosy strip of colour present along lateral line. Dorsal spots large and primarily away from ridge, between tops of parr marks. Dorsal parr marks may be present along dorsal ridge ahead of dorsal fin and number roughly five (but up to 10). Dorsal fin has black leading edge and white tip with black underneath. Anal fin has sweeping orange section. Adipose fin opaque, variable in colour and has one or two partial black lines along edge. Other fins partially orange. *Typical dead fish diagnostics: 17–21 short, blunt rakers on first gill arch, 30–70 pyloric caeca, 8–12 anal fin rays, 10–15 branchiostegal rays, no basibranchial teeth on base of tongue.*

BEHAVIOUR AND HABITAT

Primarily found in clear, low-velocity sections of large tributaries or main stems with high structural complexity (such as large woody debris, undercut banks). Like a coastal cutthroat fry, tends to occupy riffles and glides rather than pools, leaving those to aggressive coho. Stays close to the streambed, preying on bottom fauna. Second only to coho fry in aggressive and territorial tendencies. Like coho, individuals unable to establish a feeding territory will travel downstream to find other opportunities but will then have a lower chance of survival.

SIMILAR SPECIES

Very challenging to differentiate from a coastal cutthroat fry (or cutthroat-rainbow hybrids) without genetic tools. In some regions, a steelhead fry has five or more dorsal parr marks along the ridge ahead of the dorsal fin, which can be used to distinguish from a coastal cutthroat fry in these populations.

Steelhead Parr

A curious and colourful fish with a boldness not shown by other juvenile salmon rearing in freshwater

KEY CHARACTERISTICS

Snout relatively short and rounded from side. Maxillary does not reach rear margin of eye. Parr marks regular in height, straddling lateral line and mostly longer than vertical diameter of eye. Rosy strip of colour present along lateral line. Flank spotted, with some spots below lateral line. Dorsal spots are dark, medium-sized, leopard-like and away from dorsal ridge. Dorsal parr marks may be present along dorsal ridge ahead of dorsal fin but faded and usually number three to five (but up to 10). Dorsal fin spotted black with white tip that reaches three to five fin ray interspaces. Anal fin partially orange with white tip. Adipose fin opaque, variable in colour, and has one or two partial black lines along edge. Other fins partially orange. Intermediate characteristics (such as faint throat slashes) may indicate a cutthroat-rainbow hybrid, which can be relatively common in streams with both species but genetic tools are needed to confirm. *Typical dead fish diagnostics: 17–21 short, blunt rakers on first gill arch, 30–70 pyloric caeca, 8–12 anal fin rays, 10–15 branchiostegal rays, no basibranchial teeth on base of tongue.*

BEHAVIOUR AND HABITAT

Can inhabit higher velocity river reaches than during the fry life stage, but still use features like logs and pools that provide areas to hold onto or hide within. As a parr transitions from a bottom-dwelling fry, it gradually explores closer to the surface to feed on insects and drifting zooplankton and becomes more nocturnal when feeding. Will eventually maintain position in the middle of the water column rather than holding closer to the bottom. Often migrates downriver to main stem or large tributaries even when not displaced by aggression from other parr. Typically solitary, but in excellent feeding conditions can form small schools.

SIMILAR SPECIES

Distinguished from a coastal cutthroat parr by its lack of throat slashes and short maxillary, which does not reach the rear margin of the eye.

Coastal Cutthroat Fry

A flighty bottom-dweller, which seems constantly on edge, darting between hiding places

KEY CHARACTERISTICS

Maxillary sometimes reaches or extends past rear margin of eye, especially at larger sizes (>6 cm or 2.5 in). Larger individuals (>6 cm or 2.5 in) may have yellow, orange or red slashes under lower jaw. Parr marks regular in height, straddling lateral line and mostly longer than vertical diameter of eye. Rosy strip of colour often present along lateral line. Dorsal spots large and primarily away from dorsal ridge, between tops of parr marks. Dorsal parr marks rarely present along dorsal ridge ahead of dorsal fin, but when present number fewer than five. Dorsal fin has black leading edge and white tip with black underneath and occasional spots. Anal fin has sweeping orange section. Adipose fin opaque, variable in colour and has one or two partial black lines along edge. Other fins partially orange. *Typical dead fish diagnostics: 15–21 short, blunt rakers on first gill arch, 25–55 pyloric caeca, 8–12 anal fin rays, 9–12 branchiostegal rays, basibranchial teeth present on base of tongue.*

BEHAVIOUR AND HABITAT

Abundant in small streams but can be found throughout the watershed of most coastal rivers, including upper reaches of small tributaries and (short-lived) off-channel or swampy habitats. Usually dwells along stream margins in riffles and glides rather than pools, which coho tend to dominate. Darts between hiding places on the streambed in a skittish manner feeding on benthic and some drifting prey.

SIMILAR SPECIES

Very difficult to differentiate from a steelhead fry (or cutthroat-rainbow hybrids) without genetic tools since the diagnostic maxillary reaches or extends past the rear margin of the eye only sometimes. Diagnostic slashes under the jaw only occasionally present. In some regions, a steelhead fry has five or more dorsal parr marks along the ridge ahead of the dorsal fin, whereas a coastal cutthroat usually has none or, when present, fewer than five.

Coastal Cutthroat Parr

A gorgeous fish with bright colours, heavy spotting and vibrant throat slashes

KEY CHARACTERISTICS

Snout rather long and pointed from side. Maxillary reaches or extends past rear margin of eye. Yellow, orange or red slashes under lower jaw. Parr marks regular in height, straddling lateral line and mostly longer than vertical diameter of eye. Rosy strip of colour present along lateral line. Heavy spotting on flank, with many below lateral line. Dorsal spots are dark, medium-sized, leopard-like and away from dorsal ridge. Dorsal parr marks rarely present along dorsal ridge ahead of dorsal fin, but when present number fewer than five. Dorsal fin spotted black with white tip that reaches one to three fin ray interspaces. Anal fin mostly orange with white tip. Adipose fin opaque, variable in colour and has one or two partial black lines along edge. Other fins partially or mostly orange. Intermediate characteristics may indicate a cutthroat-rainbow hybrid, but genetic tools are needed to confirm. *Typical dead fish diagnostics: 15–21 short, blunt rakers on first gill arch, 25–55 pyloric caeca, 8–12 anal fin rays, 9–12 branchiostegal rays, basibranchial teeth present on base of tongue.*

BEHAVIOUR AND HABITAT

Widely distributed throughout the watershed, from the main stem all the way to ephemeral swamps in upper reaches of small tributaries. Highly mobile compared to cutthroat fry, primarily downstream but sometimes upstream. Larger individuals preferentially occupy pools when coho not present, but otherwise occupy riffles and glides. Has a varied diet that includes everything from benthic prey to surface insects and smaller fishes, allowing for opportunistic feeding.

SIMILAR SPECIES

Distinguishable from a steelhead parr by its slashes under the jaw and maxillary that reaches or extends past the rear margin of the eye.

SMOLTS

The smolt life stage marks the extraordinary transition that young salmon must make from living in freshwater to living in saltwater. Fewer than 1% of fish species undertake this difficult transition, but for salmon the effort is rewarded with bountiful feeding opportunities that the ocean provides. Salmon prepare for marine life through the process of smoltification, which is timed to coincide with the downstream migration that young salmon make towards the ocean. Smoltification involves sweeping changes to the body shape, behaviour and physio-logy of salmon.

Fry and smolts of the same species differ so greatly that early biologists assumed they were entirely different species of fish—coastal Indigenous Peoples had, of course, long recognized these as separate life stages of the same species. As young salmon begin preparing for their journey downriver, they shed their cryptic colouration that helped keep them camouflaged in the bottom of streams and lakes. Their colours and parr marks become muted as they develop a silver, shiny complexion more suited to life in open water. Their bodies elongate and their snouts become pointier making them more streamlined to improve swimming efficiency. A few species begin growing scales while smolting, but scale development is not intrinsically linked to smoltification—some species gain scales months or years before becoming smolts (coastal cutthroat and steelhead) while others develop scales after becoming smolts (pink and chum).

Smoltification is accompanied by several behavioural changes in salmon. While fry and parr typically stay near the bottom of their stream or lake, smolts must eventually migrate nearer to the surface. When smolting, salmon stop orienting themselves against the current and lose some of the territoriality of their earlier youth, instead opting to shoal together for defensive advantage against predators. When the time is right to migrate downriver (often under the cover of darkness and when the river is swollen with rain or melting snow) some species and populations will leave in massive schools to overwhelm the predators waiting for them and limit the number of casualties.

As their bodies and behaviours change to support the transition from bottom-dwelling fish to open-water migrants, smolting salmon undergo a drastic physiological transformation. The main organ involved in this change is the gill. In saltwater, fish have lower concentrations of salt in their bodies than the surrounding environment, so without regulation they would gain too much salt and lose water to the environment. During smoltification, gill proteins responsible for maintaining the salt balance are increased and work in overdrive to remove excess salt from the body. At the same time, the kidney begins producing highly concentrated urine to flush out more salt while retaining as much water as possible.

For most species, the beginning of smoltification is triggered by day length and growth rate, and the process typically takes months to fully complete. Salmon reach their peak salt tolerance near the end of the smolting process, and they typically time their downstream migration and seawater entry to coincide with this "smolt window." Fish that don't enter the marine environment during this window can still do so later, but smolting salmon that are prevented from reaching saltwater by physical migration barriers can de-smoltify and reverse many of the changes from smoltification.

As with everything about the salmon life cycle, there are no hard rules regarding the timing of smoltification that hold true across all species and populations. Chinook, coho and sockeye usually begin smolting after a year in freshwater, but sometimes in their first year or not until their third year or even later. Steelhead and coastal cutthroat can take up to seven years before smolting and migrating to the ocean, but coastal cutthroat typically smolt after two to four years and steelhead after one to three years. Pink and chum, however, leave their natal rivers immediately after gravel emergence, so the distinctions between fry, smolt and post-smolt are difficult to define. Until they begin developing scales after a month in the ocean, pink and chum look nearly the same as they did in freshwater. For the purposes of this book, we present pink and chum smolts as scaled fish in the marine environment that are roughly 6–8 cm (2.5–3 in) in fork length, since this is the small size window in which chum lose their parr marks and the spotting along their backs. The other five species are shown here as smolts during their downstream migration. The Chinook, coho and sockeye illustrations represent fish 6–8 cm (2.5–3 in) in length, whereas the coastal cutthroat and steelhead illustrated here are 15–18 cm (6–7 in) in length because they enter the marine environment as older and larger individuals.

Smolts typically migrate to the ocean between February and May, with coastal cutthroat and steelhead peaking later in the season than the other five species. Populations at southern latitudes typically migrate earlier than those farther north. Unlike post-smolts, smolts are not often seen feeding or milling in large groups at the surface. While smolts do eat during their downstream migration (mainly zooplankton and insects, as well as some fishes for coastal cutthroat and steelhead), feeding is typically more opportunistic and less noticeable than in the marine environment. Despite their vast numbers, smolts are often on the move at depth or, when surface oriented, in turbid water, making them less noticeable than other life stages. Scientists capture smolts using a variety of trapping and netting techniques, or sometimes via (non-lethal) electrofishing in small creeks.

While identification among smolts is not as straightforward as it is for the fry stage, it is generally not too challenging after some practice, with one obvious exception: Chinook and coho smolts are notoriously difficult to differentiate. In a few river systems, their traits overlap so much that the only truly reliable way to distinguish between the two species is through genetic analysis. In most locations, however, you can usually get it right after seeing enough fish from the same river system.

SUMMARY TABLE OF SMOLT KEY CHARACTERISTICS

Feature	Pink	Chum	Sockeye
Habitat	marine	marine	usually fresh or brackish water
Eye size relative to head	medium	medium	very large
Parr marks	none	present but faded; short and mostly above lateral line; usually half to two-thirds the width of interspaces	present but slightly faded; irregular, but some straddle lateral line; roughly half as wide as interspaces
Dorsal spotting	none	medium density; muted; variable in size and shape	usually none close to ridge, but has row of large, faded spots just above spaces between parr marks
Notable caudal fin characteristics	no distinct delineated black regions; gradual gradient of increasing blackness that does not reach the fork or trailing edge	distinct black region that runs parallel to the trailing edge of the fin and that reaches the fork and lobe tips	long, flowing lobes and absent of any colour other than black
Other key features	base of caudal fin has wide turquoise or green arc; scales very small	base of caudal fin has turquoise or green arc	all fins and base of caudal fin absent of any colour
Most similar species	chum	pink	chum

Coho	Chinook	Steelhead	Coastal Cutthroat
usually fresh or brackish water	usually fresh or brackish water	fresh or brackish water	fresh or brackish water
large	large	small	small
present but slightly faded; long and straddle lateral line; some are usually narrower than interspaces; often thin and oval-shaped	present but slightly faded; long and straddle lateral line; all are usually wider than interspaces; often thick and oval-shaped	present but faded; short and straddle lateral line; variable in spacing	present but faded; short and straddle lateral line; variable in spacing
medium density; shaped like circles or smudges	medium density; shaped like circles and smudges	high density; large and dark; irregularly shaped	high density; large and dark; irregularly shaped
none (inconsistent)	none (inconsistent)	shallow fork and spotted	shallow fork and spotted
snout from side short and blunt (unlike Chinook); anal fin typically has leading edge at least as long as trailing edge (unlike Chinook) 	snout from side rather pointy (unlike coho); anal fin typically has longer trailing edge than leading edge (unlike coho) 	snout from side short and rounded (unlike coastal cutthroat); flank, dorsal fin and caudal fin spotted; maxillary does not reach rear margin of eye (unlike coastal cutthroat)	snout from side rather long and pointy (unlike steelhead); flank, dorsal fin and caudal fin spotted; maxillary usually extends to or past rear margin of eye (unlike steelhead); usually has faint yellow, orange or red slashes under jaw
Chinook	coho	coastal cutthroat	steelhead

Pink Smolt

A clean, silver-sided fish with tiny scales

KEY CHARACTERISTICS

Maxillary does not reach the rear margin of the eye. Medium-sized eye relative to the size of the head. No parr marks. No spotting of any sort. Scales and scale pockets (in descaled fish) notably small. Dorsal surface gains dark-blue patches when fish is stressed from handling. Pectoral, pelvic and anal fins lack colour. Base of the caudal fin has a wide arc of turquoise or green that traces the interface of the caudal peduncle and caudal fin. Caudal fin lacks distinct black regions, instead showing a gradient in colour from green or blue to black. Black colour does not reach the fork or trailing edge of caudal fin and is concentrated in the rays rather than the spaces between them. *Typical dead fish diagnostics: 27–33 long, slender rakers on first gill arch, 130–195 long, slender pyloric caeca, 14–16 anal fin rays, 11–14 branchiostegal rays.*

BEHAVIOUR AND HABITAT

Typically found in abundant, slow-moving schools with co-migrating chum in the coastal marine environment. Habitat preference between the shallow, nearshore inclinations of pink fry and the slightly deeper preferences of pink post-smolts. Typically travels within 3 m (10 ft) of the surface during daytime, unless startled making them move deeper in the water. Surface activity ranges from raindrop-like dimples to small leaps fully out of the water. Travels day and night. Energetic when captured compared to co-migrant species.

SIMILAR SPECIES

Can sometimes be confused with co-migrating chum, but a pink smolt has no parr marks, much smaller scales and a caudal fin with a gradient of black colour rather than a distinct black region. Pink smolts begin scaling later than chum smolts.

Chum Smolt

A sturdy and busy-looking fish, especially in comparison to its pink co-migrants

KEY CHARACTERISTICS

Maxillary does not usually reach rear margin of eye. Eye medium-sized eye relative to head size. Parr marks short, faded and occurring mostly above lateral line. Parr-mark width is usually half to two-thirds width of interspaces. Dorsal surface covered in medium-density, muted spots that are variable in size and shape and are visible from side. Base of caudal fin has an arc of turquoise or green that traces interface of caudal peduncle and caudal fin. Caudal fin has a distinct black region with leading edge running mostly parallel to trailing edge. Black colour reaches fork and trailing edge of caudal fin and is strongly contrasted with rest of fin. *Typical dead fish diagnostics: 20–26 short, blunt rakers on first gill arch, 160–185 pyloric caeca, 13–14 anal fin rays, 13–14 branchiostegal rays.*

BEHAVIOUR AND HABITAT

Forms schools with co-migrating pinks in the coastal marine environment. Like pinks, tends to occupy an intermediary depth and distance from shore compared to the younger fry and older post-smolt life stages, but normally still within 3 m (10 ft) of the surface and within a stone's throw of the shore. Often found milling in small, sheltered coves in schools identifiable by raindrop-like dimpling and occasional leaping fish. Travels day and night.

SIMILAR SPECIES

Once a chum smolt loses its parr marks, it can be differentiated from a pink smolt by its larger scales and the distinct black region in its caudal fin. Chum smolts begin scaling earlier than pink smolts.

Sockeye Smolt

A uniquely attentive looking fish with its huge eye and docile nature

KEY CHARACTERISTICS

Maxillary does not reach rear margin of eye. Eye large relative to head size. Parr-mark width is roughly half the width of interspaces. Slightly faded "piano key" parr marks that straddle lateral line and can be variable in length and position. Dorsal surface usually lacks spots near dorsal ridge but with a row of large, faded spots just above spaces between parr marks. Fins and skin at base of caudal fin do not have any colour other than black. Caudal fin has a deep fork and long flowy lobes with no distinct black regions. Black colour in caudal fin concentrated in rays rather than spaces between them. Fragile when handled. *Typical dead fish diagnostics: 32–37 long, slender, close-set rakers on first gill arch, 65–95 pyloric caeca, 14–16 anal fin rays, 13–15 branchiostegal rays.*

BEHAVIOUR AND HABITAT

Primarily found in reaches of rivers directly between the rearing lake and the ocean. Migrates from lake to ocean in large schools that move in coordinated pulses at night to overwhelm predators and reduce predation risk. Downstream migration often slows or stops during the day, especially in less turbid sections of the river (with clearer water) where predators can see better. Tends to travel near the surface in shallow rivers and midwater in deeper rivers, facing downstream and swimming faster than the current. For many populations, especially those in single-lake systems, the majority of the smolt exodus can occur in just a few days. Exhibits little surface activity.

SIMILAR SPECIES

A sockeye smolt can look like an equivalent-sized chum smolt but has a larger eye and lacks colouration in and around the caudal fin.

Coho Smolt

A stubby-nosed fish in an awkward adolescent phase

KEY CHARACTERISTICS

Snout short and blunt from side. Maxillary ordinarily just reaches rear margin of eye. Eye large relative to head size. Parr marks are slightly faded, straddle lateral line and are usually long, thin and oval. Some parr marks are usually narrower than interspaces. Dorsal spots are medium-density circles or smudges irregular in shape and spacing and visible from side. Anal fin is often sickle-shaped with a black smudge and/or white tip, and usually has a white leading edge at least as long as trailing edge. Adipose fin fully pigmented. Caudal fin colouration inconsistent. Overall colour variable and dependent on habitat and age at migration. *Typical dead fish diagnostics: 19–27 short rakers on first gill arch, 50–85 pyloric caeca, 13–14 anal fin rays, 13–14 branchiostegal rays, much smaller otoliths than similar-sized Chinook.*

BEHAVIOUR AND HABITAT

Can be found throughout the watershed, from shallow headwater streams to estuaries. Habitat choice is dependent on age at migration, which is highly variable—many coho fry migrate to the estuary where they then mature into smolts. Stays under cover during the day and becomes active at night. Displays less territoriality than during the fry life stage, allowing for schooling to occur (but not normally in the large aggregations exhibited by pink, chum and sockeye). Depth preference variable, but often nearer the surface than sockeye and Chinook. Exhibits little surface activity.

SIMILAR SPECIES

A coho smolt is easily confused with a Chinook, but has a short, blunt snout and usually has a longer leading edge than trailing edge of the anal fin.

Chinook Smolt

A thickset migrant with a football-shaped snout

KEY CHARACTERISTICS

Snout rather long and pointy (or "football-shaped") from side. Maxillary ordinarily just reaches rear margin of eye. Eye large relative to head size. Parr marks are slightly faded, straddle lateral line and are usually long, thick and oval. Parr marks are usually wider than interspaces. Dorsal spots are medium-density circles or smudges irregular in shape and spacing and are visible from side. Anal fin usually has a longer trailing edge than leading edge and can occasionally have a white leading edge and/or tip. Adipose fin has a dark outer edge and often a diagnostic clear, unpigmented centre. Caudal fin colouration inconsistent. Overall colour variable and dependent on habitat. Deep bodied. *Typical dead fish diagnostics: 20–25 short rakers on first gill arch, 140–185 pyloric caeca, 15–19 anal fin rays, 16–18 branchiostegal rays, much larger otoliths than similar-sized coho.*

BEHAVIOUR AND HABITAT

Mainly found in tributaries and main stems of large streams and rivers. More active at night than daytime during downstream migration, but that preference disappears closer to the river mouth. Forms schools but not in large aggregations. Rarely travels near the surface, instead preferring the middle and bottom of the water column. Exhibits little surface activity.

SIMILAR SPECIES

A notorious look-alike to a coho smolt, a Chinook smolt can usually be differentiated by its long, pointed ("football-shaped") snout, deeper body and typically a longer trailing edge than leading edge of the anal fin.

Steelhead Smolt

A hardy smolt with ample spots, short snout and power to spare

KEY CHARACTERISTICS

Snout relatively short and rounded from side. Snout, eye and cheek often spotted. Maxillary does not reach rear margin of eye. Eye small relative to head size. Faded parr marks are short, variably spaced and straddle lateral line. Rosy strip of colour often present along lateral line. Large, dense, dark leopard-like spots on dorsal surface and flank, including some below lateral line. Dorsal spotting sparse along dorsal ridge ahead of dorsal fin. Dorsal and adipose fins spotted. Pelvic and anal fins variable in colour but often have white tips. Caudal peduncle thick. Caudal fin typically spotted with a shallow fork. Intermediate characteristics may indicate a cutthroat-rainbow hybrid, but genetic tools are needed to confirm. *Typical dead fish diagnostics: 17–21 short, blunt rakers on first gill arch, 30–70 pyloric caeca, 8–12 anal fin rays, 10–15 branchiostegal rays, no basibranchial teeth on base of tongue.*

BEHAVIOUR AND HABITAT

Once smolted, mainly found in tributaries and main stems of rivers. Unlike other species, exhibits little preference for nocturnal movements except in upper reaches of rivers. More solitary than other salmon species but will form small schools in systems with larger populations. Migrates slowly relative to other species, often with lengthy stopovers. Typically travels in the middle or bottom of the water column. Opportunistically feeds during migration. Exhibits little surface activity.

SIMILAR SPECIES

A steelhead smolt has a shorter, rounder snout than a coastal cutthroat smolt, less spotting below the lateral line and a maxillary that does not reach the rear margin of the eye.

Coastal Cutthroat Smolt

A long-jawed and heavily spotted fish whose thickness and iridescence makes it seem particularly three dimensional as it silvers up for the ocean

KEY CHARACTERISTICS

Snout rather long and pointed from side. Snout, eye and cheek often spotted. Maxillary reaches or extends past rear margin of eye. Eye small relative to head size. Usually has diagnostic faint yellow, orange or red slashes under lower jaw, which often disappear in marine environment. Parr marks faded, short, variably spaced and straddle lateral line. Large, dense, dark leopard-like spots on dorsal surface and flank, including many below lateral line. Dorsal and adipose fins spotted. Pelvic and anal fins variable in colour but often have white tips. Caudal peduncle thick. Caudal fin typically spotted and with a shallow fork. Intermediate characteristics may indicate a cutthroat-rainbow hybrid, but genetic tools are needed to confirm. *Typical dead fish diagnostics: 15–21 short, blunt rakers on first gill arch, 25–55 pyloric caeca, 8–12 anal fin rays, 9–12 branchiostegal rays, basibranchial teeth present on base of tongue.*

BEHAVIOUR AND HABITAT

Once smolted, mainly found in tributaries and main stems of rivers, as well as estuaries. Migrates quickly downstream, coordinating with out-going tides in lower reaches of large rivers. More solitary than other salmon species but will form small schools in systems with larger populations. Typically travels in the middle or bottom of the water column. Opportunistically feeds during migration. Exhibits little surface activity but can occasionally be seen feeding on insects or fish near or at the surface.

SIMILAR SPECIES

The throat slashes of the coastal cutthroat smolt are diagnostic when present. When lacking slashes, a pointier snout, heavier spotting below the lateral line and maxillary length reaching or extending past the rear margin of the eye distinguish a cutthroat smolt from a steelhead.

POST-SMOLTS

Post-smolt is a somewhat nebulous term that refers to the early marine life of salmon. For all but coastal cutthroat trout, this is a period of drastic change during which most salmon undertake a coastal migration from their natal estuary to the open ocean, where they'll spend the next one to eight years of their lives.

The beginning and end of this life stage are not well-defined across all species. The post-smolt stage is often considered to be the period when salmon are rapidly growing in saltwater following smoltification and prior to maturing into the adult phase. If applied across all species, however, this definition would include fish so small they still lacked scales. Most salmon migrate to sea at approximately 8 cm (3 in) or longer, but there are some exceptions—pink, chum and subsets of the other species enter saltwater at smaller sizes and only reach 8 cm (3 in) in length after one to two months in saltwater. Because of these exceptions, it is useful for identification purposes to define the post-smolt stage by body size.

In this book, we consider salmon to be post-smolts after they enter the ocean and are between 10 cm (4 in) and 25 cm (10 in) in fork length. This upper size limit for the post-smolt stage is arbitrary. In practice, defining the endpoint for the post-smolt life stage is less important than defining the beginning, since most species move offshore just a few months after emerging from freshwater and are therefore not seen during the transition from post-smolt to adulthood by any but the most inquisitive humans. Coastal cutthroat trout typically do not travel far from their natal streams, and post-smolt and ocean adult coastal cutthroat look similar. Chinook and coho salmon can overwinter in coastal systems (such as the Salish Sea) in their first year at sea, but by 25 cm (10 in) these fish generally resemble something in-between a post-smolt and an adult.

The post-smolts illustrated in this book all range 10–12 cm (4–5 in) in length except for steelhead and coastal cutthroat, which range 20–23 cm (8–9 in) in length because these fish typically reach the ocean at lengths greater than 16 cm (6 in). Pink and chum post-smolts are in their first year of life, while sockeye, coho and stream-type Chinook post-smolts are usually in their second year. Coastal cutthroat and steelhead post-smolts migrate to sea at any time between their first to sixth years of life but most likely at two to three years of age.

Early marine life is a difficult time for salmon. After enduring the physiological stress of adapting to saltwater, they must then commence a long migration filled with predators and other threats such as disease and competition for food. The post-smolt life stage is consequently a period of extremely high mortality; for most salmon species, 50–90% die as post-smolts.

Salmon post-smolts have a diverse diet of zooplankton, insects and smaller fishes. Pink and chum are largely planktivorous at this life stage, and sockeye eat mostly zooplankton throughout their entire life cycle, but in general as post-smolts grow so does their reliance on fish prey. Post-smolts are themselves an important food source for seabirds, fish and pinnipeds such as sea lions.

Post-smolts can be seen between March and August along the North Pacific coastline during their (generally northward) migration along the continental shelf before moving to offshore waters. Specific timing depends on the species and location, but peak abundances often occur in May in Washington, June in British Columbia and July in Alaska. At these peak times, schools of fish can become indistinguishable, and instead, observers are occasionally treated to a never-ending parade of migrating salmon exhibiting surface activity. Unlike most other nearshore marine fishes, many species of post-smolt salmon leap out of the water (which, as it happens, is an effective way for them to remove ectoparasites). Some can be found feeding or milling on calm days when their dorsal surfaces break the water to form a distinctive dimpling pattern. On a sunny day, keen-eyed observers on boats can see the torpedo-like bodies of post-smolts underwater that dart in cohesive schools.

Few people other than scientists interact with post-smolts directly. Even then, we know relatively little about this life stage, and it's only in the past decade or two that researchers have started focusing on it. Now that we know how important early marine life is to salmon, more research programs are being implemented to better understand the ecology of post-smolts. These fish can be captured by trawlers or purse seiners using nets with fine mesh sizes or by smaller purse seines pulled in by hand from powerboats. Larger coho and Chinook post-smolts can be caught with microtrolling techniques (trout-sized hook-and-line gear attached to downriggers). Post-smolts are less delicate than fry and smolts but care still needs to be taken when handling them; most species change colour and gain blue splotches on their dorsal surfaces when stressed at this life stage.

This may be the most difficult life stage to distinguish species because they all look similar and silvery at first glance. By the time salmon are post-smolts, their parr marks have faded or fully disappeared and most species are roughly the same size at a given time and location. But from this similarity comes opportunity. Post-smolts provide the chance for observers to hone their identification skills and appreciate the minute details that characterize these beautiful fish. In contrast to other life stages, caudal fin patterning is probably the most important feature for post-smolt identification, and other under-used identifying characteristics like top-down snout and body shape are remarkably diagnostic for these fish.

SUMMARY TABLE OF POST-SMOLT KEY CHARACTERISTICS

Feature	Pink	Chum	Sockeye
Caudal fin	no distinctly delineated black regions; gradual gradient of increasing blackness that does not reach fork or trailing edge; iridescent flashes of green, blue or gold when in water 	distinct black region that runs parallel to trailing edge of fin and that reaches fork and lobe tips; iridescent flashes of green, blue or gold when in water 	two distinct black regions concentrated in long, flowy lobes, with heavy contrast to rest of fin; absent of any colour 
Eye size relative to head	medium to large	small	large
Parr marks	none	none	present but fade as fish grows; rarely seen on fish larger than 11 cm (4.5 in)
Dorsal spotting	low-density, medium-sized, muted ovals or "kidney beans"	high-density, fine, muted "pinpricks"	variable in density and shape, but when present medium-sized and muted
Snout shape from above	long and tapered 	long and tapered 	rounded 
Body shape from above (from operculum to caudal fin)	equally wide from operculum to halfway between dorsal fin and operculum, slight taper thereafter; steep taper between adipose fin and caudal fin 	equally wide from operculum to leading edge of dorsal fin, tapering gradually to caudal fin 	widest at operculum and tapering gradually to caudal fin 
Other key features	base of caudal fin has wide turquoise or green arc; scales very small	base of caudal fin has turquoise or green arc	all fins and base of caudal fin absent of any colour (other than black)
Most similar species	chum	pink	pink and chum

Coho	Chinook	Steelhead	Coastal Cutthroat
distinct black region that usually reaches fork and lobe tips and that can extend along ventral and/or dorsal edge; iridescent flashes of green, blue, gold or purple when in water 	distinct black region(s) that can extend along ventral and/or dorsal edge, with minimal black at fork and lobe tips; iridescent flashes of green, blue, gold, purple when in water 	black region near trailing edge that is variable in size, intensity and distinctness; shallow fork and black spotting 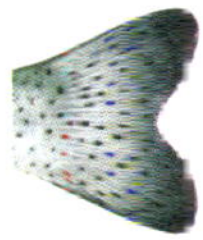	no black region; very shallow fork and black spotting; overall colour initially variable but gradually becomes yellow-brown as fish grows 
medium	medium to large	medium	medium
present but fade as fish grows; rarely seen on fish larger than 15 cm (6 in)	present but fade as fish grows; rarely seen on fish larger than 15 cm (6 in)	present but fade as fish grows; rarely seen on fish larger than 20 cm (8 in)	present but fade as fish grows; rarely seen on fish larger than 20 cm (8 in)
medium-density, large, dark circles and smudges	medium-density, large, dark circles and smudges	high-density, large, dark, leopard-like spots	high-density, large, dark, leopard-like spots
short and blunt 	long and rounded 	inconsistent	inconsistent
widest at operculum before tapering gradually to caudal fin; wide overall 	widest at operculum before tapering gradually to caudal fin; wide overall 	equally wide from operculum to leading edge of dorsal fin, tapering gradually to caudal fin; wide overall 	equally wide from operculum to leading edge of dorsal fin. tapering gradually to caudal fin; wide overall 
snout from side short and blunt (see smolt section); anal fin often has black smudge and/or white tip and usually has white leading edge at least as long as the trailing edge (unlike Chinook) 	snout from side rather pointy (see smolt section); anal fin generally lacks colour but can occasionally have white leading edge and/or tip and usually has longer trailing edge than leading edge (unlike coho) 	snout from side short and rounded (unlike coastal cutthroat); flank and dorsal fin spotted; maxillary does not reach rear margin of eye (unlike coastal cutthroat)	snout from side rather long and pointy (unlike steelhead); flank and dorsal fin spotted; maxillary usually extends to or past rear margin of eye (unlike steelhead); pectoral, pelvic and anal fins often partially yellow or orange, especially in larger (>20 cm or 8 in) individuals
Chinook	coho	coastal cutthroat	steelhead

Pink Post-Smolt

An energetic and inquisitive fish that sports small scales and distinctive spots shaped like ovals or kidney beans

KEY CHARACTERISTICS

Snout long and tapered from above. Maxillary ordinarily does not reach rear margin of eye. Eye medium-large relative to head size. Parr marks never present. Scales and scale pockets (in descaled fish) notably small. Dorsal surface colour generally army green but can range to aquamarine. Dorsal markings characteristically regularly spaced, large, muted spots shaped like ovals or kidney beans. Dorsal spots visible from the side. Base of caudal fin has a wide arc of turquoise or green that traces interface of caudal peduncle and caudal fin. Caudal fin lacks distinct black regions, instead showing a gradient in colour from green or blue to black. Black colour does not reach fork or trailing edge of caudal fin and is concentrated in rays rather than spaces between them. Caudal fin and base of caudal fin flashes green, blue or gold in water. From above, body is equally wide from operculum to halfway between operculum and dorsal fin, whereafter it tapers slightly in width to adipose fin and greatly from there to caudal fin. *Typical dead fish diagnostics: 27–33 long, slender rakers on first gill arch, 130–195 long, slender pyloric caeca, 14–16 anal fin rays, 11–14 branchiostegal rays.*

BEHAVIOUR AND HABITAT

Almost always in mixed-species schools with similar-sized post-smolts, but most often with chum as both migrate from freshwater as fry and grow rapidly during their coastal outmigration together. Variable habitat and depth preference, but commonly found in the nearshore environment slightly deeper than pink and chum fry. Frequently leaps fully out of the water. When captured, swims vigorously into barriers, shedding small scales before acclimatizing.

SIMILAR SPECIES

The chum post-smolt is similar, but a pink post-smolt has bigger eyes, smaller scales, larger dorsal spots and a caudal fin with a gradient of black colour rather than a distinct black region.

Chum Post-Smolt

A robust-looking fish with a stout frame and small eyes

KEY CHARACTERISTICS

Snout long and tapered from above, often with a dark leading edge. Maxillary ordinarily does not reach rear margin of eye. Eye small relative to head size. Parr marks from fry stage fully disappeared. Dorsal surface generally green-gold or olive-brown and marked by high-density pinprick spotting that is not visible from side. Base of caudal fin has an arc of turquoise or green that traces interface of caudal peduncle and caudal fin. Caudal fin has a distinct black region with a leading edge that runs mostly parallel to trailing edge of fin. Black colour reaches fork and trailing edge of caudal fin and is strongly contrasted with rest of fin. Caudal fin and base of caudal fin flashes green, blue or gold in water. From above, body is equally wide between operculum and leading edge of dorsal fin, after which its width tapers gradually towards caudal fin. *Typical dead fish diagnostics: 20–26 short, blunt rakers on first gill arch, 160–185 pyloric caeca, 13–14 anal fin rays, 13–14 branchiostegal rays.*

BEHAVIOUR AND HABITAT

Forms mixed-species schools with other post-smolts, particularly pink and sockeye. Found in a variety of habitats and depths but is a frequent visitor within 100 m (330 ft) of the shoreline. Frequently leaps fully out of the water. Less energetic than pink post-smolts when captured.

SIMILAR SPECIES

A chum post-smolt can be differentiated from a pink post-smolt by its relatively small eye, normal-sized scales, pinprick dorsal spotting and distinct black region in its caudal fin.

Sockeye Post-Smolt

A striking fish with large eyes and flowing caudal fin

KEY CHARACTERISTICS

Snout rounded from above. Maxillary does not reach rear margin of eye. Eye large relative to head size. Smaller individuals (<11 cm or 4.5 in) often have faded parr marks. Dorsal surface highly variable, with colour ranging from aquamarine to purple and with spotting from absent to dense. Dorsal spots, when present, medium-sized, muted in colour and difficult to observe from side. Fins and skin at base of caudal fin do not have any colour other than black. Caudal fin deeply forked with long, flowy lobes that have two distinct black regions that contrast strongly with rest of fin, are thickest at lobe tips and do not quite reach trailing edge of fin. From above, widest part of body is at operculum, after which body width tapers consistently towards caudal fin. Clean look. *Typical dead fish diagnostics: 32–37 long, slender, close-set rakers on first gill arch, 65–95 pyloric caeca, 14–16 anal fin rays, 13–15 branchiostegal rays.*

BEHAVIOUR AND HABITAT

Usually schools with other species, especially pink and chum post-smolts, but population-specific differences in sockeye migration timing makes them more likely than pink and chum to form single-species schools. Often found feeding on zooplankton or holding in the upper 10 m (30 ft) of nearshore waters less than 50 m (160 ft) in depth. Migrates quickly, and when travelling will move deeper and farther away from shore. Looks like a lazy jumper when breaking the surface of the water as it tends to slide its tail along the surface rather than fully leaping. One of the more docile and delicate post-smolt species when captured.

SIMILAR SPECIES

Pink and chum post-smolts can be somewhat similar, but a sockeye smolt has larger eyes and is the only species at this life stage with no colour other than black in all its fins and the base of its caudal fin.

Coho Post-Smolt

A daunting predator of smaller fish with bull-like nostrils and a blunt snout

KEY CHARACTERISTICS

Snout short and blunt from above and side. Maxillary reaches or extends past rear margin of eye. Eye medium sized relative to head size. Parr marks often faded in smaller individuals (<15 cm or 6 in), with some parr marks no wider than spacing between them. Dorsal surface colour variable. Dorsal spots large, dark circles or smudges irregular in shape and spacing, easily seen from side. Anal fin often sickle-shaped with a black smudge and/or white tip, and usually with a white leading edge at least as long as trailing edge. Adipose fin always fully pigmented with a dark outer edge. Caudal fin with distinct black region that runs parallel to trailing edge near fork and widens at lobe tips. Black colour usually reaches fork and trailing edge of caudal fin and can extend along ventral and/or dorsal edge of fin. Caudal fin lobes have large patches of colour that flash green, blue, gold or purple in water. From above, wide overall with widest point at operculum and gradually tapering in width to caudal fin. *Typical dead fish diagnostics: 19–27 short rakers on first gill arch, 50–85 pyloric caeca, 13–14 anal fin rays, 13–14 branchiostegal rays, much smaller otoliths than similar-sized Chinook.*

BEHAVIOUR AND HABITAT

Mills about in coastal straits and bays during its slow migration, sometimes over winter. Occasionally does not migrate to the open ocean at all. Tends to stay in deeper water than pink, chum and sockeye post-smolts, but is still regularly captured with them in small numbers. Forms smaller schools with other coho and sometimes Chinook post-smolts. Performs exuberant leaps out of the water, occasionally by itself with no other nearby surface activity. This voracious predator can provoke schools of small forage fish to "boil" on the surface while hunting and will actively feed on pink and chum fry if captured in the same seine.

SIMILAR SPECIES

Can usually be distinguished from a Chinook post-smolt by its blunt snout (from the side), anal fin that typically has a leading edge at least as long as the trailing edge, and caudal fin that has black in the fork and lobe tips.

Chinook Post-Smolt

An eye-catching, beefy fish with a pointy snout and an aloof disposition

KEY CHARACTERISTICS

Snout long and rounded from above and rather pointed or "football-shaped" from side. Gums black in larger individuals (>15 cm or 6 in). Maxillary reaches or extends past rear margin of eye. Eye medium-large relative to head size. Parr marks often faded in smaller individuals (<15 cm or 6 in), with all parr marks usually wider than spacing between them. Dorsal surface colour usually olive-green. Dorsal spots large, dark circles or smudges irregular in shape and spacing and easily seen from side. Anal fin typically lacks colour, but occasionally has a white leading edge and/or tip, and usually has a longer trailing edge than leading edge. Adipose fin has a dark outer edge and often a diagnostic clear, unpigmented centre. Caudal fin with one or two distinct black regions concentrated in lobes; black colour can extend along the ventral and/or dorsal edge of fin but is minimal at fork and lobe tips. Caudal fin flashes green, blue, gold or purple in water. From above, wide overall with widest point at operculum and gradually tapering in width to caudal fin. Deep bodied. Often smells metallic, musky fishy or (some say) like pungent kelp. *Typical dead fish diagnostics: 20–25 short rakers on first gill arch, 140–185 pyloric caeca, 15–19 anal fin rays, 16–18 branchiostegal rays, much larger otoliths than similar-sized coho.*

BEHAVIOUR AND HABITAT

Can be found alongside other species in the nearshore environment, but typically remains deeper and farther from shore. When nearshore, regularly found in or near seagrass. Often travels mid-channel. Travels slowly along the coast, like coho, making migration stopovers that can last through winter or longer. An active hunter of smaller fishes. Typically moves slowly and near the bottom of seines when captured.

SIMILAR SPECIES

Unlike coho, the Chinook post-smolt has a rather pointy snout (from the side), an anal fin that usually has a longer trailing edge than leading edge and a caudal fin that lacks black colour at the fork and lobe tips.

Steelhead Post-Smolt

A rarely seen, speedy migrant whose rounded snout and short maxillary give it an oddly proportioned face

KEY CHARACTERISTICS

Snout relatively short and rounded from side. Snout, eye and cheek often spotted. Maxillary does not reach rear margin of eye. Eye medium-sized relative to head size. Parr marks absent in larger individuals (>20 cm or 8 in) and when present are short and faded. Dorsal surface colour usually blue-green. Dorsal surface and flank have large, dense, dark leopard-like spots, with few below lateral line. Dorsal spotting sparse along dorsal ridge ahead of dorsal fin. Dorsal and adipose fins also spotted. Pelvic and anal fins variable in colour but often have white tips. Caudal peduncle thick. Caudal fin typically spotted with a shallow fork, which has a black region near trailing edge that is variable in size, intensity and distinctness. From above, equally wide from operculum to leading edge of dorsal fin before tapering gradually to caudal fin; wide overall. Intermediate characteristics may indicate a cutthroat-rainbow hybrid, but genetic tools are needed to confirm. *Typical dead fish diagnostics: 17–21 short, blunt rakers on first gill arch, 30–70 pyloric caeca, 8–12 anal fin rays, 10–15 branchiostegal rays, no basibranchial teeth on base of tongue.*

BEHAVIOUR AND HABITAT

Perhaps the fastest coastal migrant of any anadromous salmonid, moving quickly offshore to the open ocean after leaving its natal rivers. Tends to travel mid-channel and at greater depth than other species. A voracious predator of fishes and invertebrates. Moves individually or in small groups, but often caught with similarly sized Chinook and coastal cutthroat post-smolts. Exhibits little surface activity. Rarely encountered outside of the first few days of coastal migration near productive steelhead river systems.

SIMILAR SPECIES

Sometimes mistaken for coastal cutthroat, the steelhead post-smolt has a shorter, rounder snout (from the side), a dark region in the caudal fin and a maxillary that does not reach the rear margin of the eye.

Coastal Cutthroat Post-Smolt

A pointy-faced predator of the super-shallow nearshore, perfectly camouflaged for seagrass beaches

KEY CHARACTERISTICS

Snout rather long and pointed. Snout, eye and cheek often spotted. Maxillary reaches or extends past rear margin of eye. Eye medium-sized relative to head size. Diagnostic faint yellow, orange or red slashes occasionally under lower jaw. Parr marks absent in larger individuals (>20 cm or 8 in) and when present are short and faded. Dorsal surface colour dependent on habitat (bluer in the ocean, yellowish green or brown at beaches or estuaries) Dorsal surface and flank have large, dense, dark leopard-like spots, extending well below lateral line. Dorsal and adipose fins also spotted. Pectoral, pelvic and anal fins often partially yellow or orange, especially in larger individuals (>20 cm or 8 in). Caudal fin has a shallow fork and is often brownish yellow and spotted, especially in larger fish (>20 cm or 8 in) found nearshore along beaches. From above, equally wide from operculum to leading edge of dorsal fin before tapering gradually to caudal fin; wide overall. Intermediate characteristics may indicate a cutthroat-rainbow hybrid, but genetic tools are needed to confirm. *Typical dead fish diagnostics: 15–21 short, blunt rakers on first gill arch, 25–55 pyloric caeca, 8–12 anal fin rays, 9–12 branchiostegal rays, basibranchial teeth present on base of tongue.*

BEHAVIOUR AND HABITAT

Has the most distinct habitat preferences of any post-smolt residing almost exclusively in shallow nearshore beaches with eelgrass or other nursery features. Stays in its natal estuary or migrates among local estuaries, but typically does not travel more than 50–70 km (30–45 mi) away from the stream of its birth. Typically follows shorelines but will occasionally cross bodies of water. Diet includes flying insects; thus, it feeds on the surface more than other species. Cryptic but will exhibit surface activity when feeding or provoked.

SIMILAR SPECIES

The steelhead post-smolt is similar, but the coastal cutthroat has a pointier snout and a maxillary that reaches or extends past the rear margin of the eye. It occasionally has diagnostic faint slashes beneath the lower jaw.

OCEAN ADULTS

The lives of Pacific salmon are guided by their need to reach the feeding opportunities provided by the ocean. Most salmon spend the majority of their lives taking advantage of the ocean's bounty, growing to much larger sizes than they could in freshwater alone. Adult salmon roam the high seas in the hundreds of millions, patrolling the surface waters of the North Pacific Ocean while some remain coastal along the continental shelf.

Salmon spend one to five years in the ocean, depending on the species and population, often travelling thousands of kilometres each year in groups. Despite ocean residency being longer than freshwater residency for most salmon, we know relatively little about their marine phase outside of what occurs in coastal waters, other than basic information on their diets and spatial distribution.

As adults, salmon have a diverse diet that is primarily zooplankton, squid and smaller fishes. Many of the differences in prey choice between salmon species are influenced by location. For example, sockeye and steelhead, which migrate farther west than the other species, consume a huge number of squid when in the central North Pacific Ocean. However, some real preferences exist. Coho, Chinook, steelhead and coastal cutthroat all eat more fish than do pink, chum and sockeye. Pink and sockeye are generalists that eat everything but with a preference for zooplankton. Chum is the only species that feeds on jellyfish in meaningful amounts.

In general, the larger that fish grow, the less vulnerable they are to being eaten by predators. This is certainly true for salmon, whose mortality rates (in the absence of fishing) decrease with age as they become too large for birds and most fishes to eat. Adult salmon in the open ocean probably have relatively few predators, with salmon sharks perhaps being the main one, but in general little is known about what eats salmon on the high seas. In coastal waters, adult salmon must contend with seals, sea lions, resident orcas (which prefer larger-bodied and farther-migrating Chinook but eat all species), more salmon sharks and, of course, humans.

While the most abundant species of salmon (pink, chum and sockeye) almost always move offshore to the Gulf of Alaska and the Bering Sea, there are exceptions among the less-abundant species. Coastal cutthroat and some coho, Chinook and steelhead remain in coastal waters throughout their adult life. Coho that move offshore generally stay within the Gulf of Alaska rather than venturing farther west. Chinook travel long distances from their home river, but rarely more than a couple hundred kilometres or over a hundred miles from shore (except in the Bering Sea), potentially because of their benthic feeding tendencies. A few steelhead do stay coastal, but most migrate rapidly offshore upon entering the ocean; indeed, steelhead have the broadest distribution of all salmon, ranging as far south as the 40th parallel and as far west as the Kamchatka Peninsula in Russia. In contrast, coastal cutthroat seldom venture farther than 50–70 km (30–45 mi) from their natal stream and occasionally never beyond their estuary.

There is no strict rule for the size or age at which salmon mature (become adults capable of reproducing), especially across species. For identification purposes, once salmon reach lengths of approximately 25 cm (10 in) in the ocean, they usually resemble something in between a post-smolt and adult. After this point, they quickly adopt most of their adult characteristics: their eyes become smaller relative to the rest of their bodies, their snouts grow pointier for greater swimming efficiency and their caudal fins lose their forks. Most salmon spend most of their lives looking like ocean adults, even if not sexually mature, not changing much until they eventually develop spawning colours, sometimes years later. Like most fish, salmon continue growing throughout their lives, and although growth rate slows with age it is still a rapid gain during peak summer months just prior to their spawning migrations.

Once salmon have sexually matured and are ready to reproduce, they begin their lengthy migration to their home rivers. Nearly all fishing for wild salmon occurs during this relatively short period of their lives. Pacific salmon can no longer be legally caught on the high seas in accordance with international agreements, yet they are heavily targeted on the coast and in the mouths of large rivers before they transition to their spawning colours and begin to lose the fat they built up in the ocean. The ocean adult life stage is therefore the one recognized by most recreational and commercial fishers and seen by consumers in grocery stores. Commercial fisheries of adult salmon use hooks and lines (suspended from long poles on trollers), gill nets (vertical rectangular nets that fish get stuck in) or purse seines (nets set around fish in a circle and cinched from the bottom to prevent escape). Recreational fishers use hook-and-line techniques, but with fishing rods on small power boats. Researchers occasionally employ surface or midwater trawls (nets suspended in the water column and towed behind a large vessel), especially in the high seas.

Distinguishing differences among species is generally much easier for adult salmon than for juveniles. While still sleek and silver like post-smolts, adults exhibit more obvious species-specific characteristics, so weighing multiple traits is often not necessary. Instead, knowing just a couple key characteristics for each species should be sufficient for adult identification. Two pairs of species can be challenging for novices to differentiate at first—Chinook and coho, and chum and sockeye—but even those are straightforward compared to their juvenile counterparts.

SUMMARY TABLE OF OCEAN ADULT KEY CHARACTERISTICS

Feature	Pink	Chum	Sockeye
Caudal fin spotting	large, dark, oval blotches on both lobes	none	almost always none (spotting rare and if present, only on top quarter of upper lobe)
Dorsal spotting	large, dark, oval spots	none, but sometimes fine black speckling	none, but sometimes fine black speckling
Colour of lower jaw gumline at base of teeth	black	grey or black	white
Colour of rest of inside of mouth	white, pink, grey and black	mostly white and pink	white, pink, grey and black
Anal fin shape	leading edge shorter than base	leading edge shorter than base	leading edge shorter than base
Typical weight at maturity	1–2.5 kg (2–5.5 lbs)	4–9 kg (9–20 lbs)	1.6–4.5 kg (3.5–10 lbs)
Typical length at maturity	35–55 cm (14–22 in)	60–85 cm (24–33 in)	45–65 cm (18–26 in)
Dorsal colour	green to blue	light or metallic blue	greenish blue to steel blue
Other key features	scales small; lateral line often pink-hued; body soft and often slimy	caudal peduncle narrow; caudal fin with silver streaks, extending from base of fin and covering roughly half of fin; pupil large; anal and pelvic fins often have prominent white tips	gums smooth; caudal fin grey throughout; body often slimy
Most similar species	small Chinook	sockeye	chum

Coho	Chinook	Steelhead	Coastal Cutthroat
sometimes none, but when present are small, dispersed, circular black spots, usually only on upper lobe	medium-sized, moderately dense, circular black spots, usually on both lobes but not always; most variable of all species	small, dense, circular black spots on both lobes, extending in rows parallel to fin rays	medium-sized, dense, circular black spots on both lobes, extending in rows parallel to fin rays
occasionally none, but usually small, circular or irregular, dispersed black spots	medium-sized, moderately dense and irregular (wavy) black spots; most variable of all species	small, dense, circular black spots, generally only above lateral line	medium-sized, dense, circular black spots, extending below lateral line
white or light grey	black	white	white
mostly black and grey	almost entirely black	almost entirely white	almost entirely white
leading edge shorter than base	leading edge shorter than base	leading edge longer than base	leading edge longer than base
2–7 kg (4.5–15 lbs)	4.5–20 kg (10–44 lbs)	1.8–7 kg (4–15 lbs)	1–2 kg (2–4.5 lbs)
45–75 cm (18–30 in)	55–115 cm (22–45 in)	45–90 cm (18–35 in)	30–50 cm (12–20 in)
blue to black	green to black	metallic blue or dark blue	yellowish green to brown
fin rays of caudal fin feel rough (like edge of dime) when lightly running fingernail along them (unlike Chinook)	fin rays of caudal fin feel smooth when lightly running fingernail along them (unlike coho); dark outline and features when underwater; ridge on upper caudal peduncle behind attachment point of outer caudal fin ray (unlike coho)	snout from side short and rounded; caudal fin has flat trailing edge; sharp transition from dorsal colour to silver/white flank	all fins partially yellow, orange or brown; spots extend below lateral line and often on and/or below eye; tip of snout often yellow around mouth
Chinook and steelhead	coho and pink	coho	unique

Pink Ocean Adult

A small, erratic and lively fish carpeted in unmistakeable oval spots and tiny scales

KEY CHARACTERISTICS

Inside of mouth white, pink, grey and black. Lower jaw has black gumline at base of small teeth. Scales small and loose. Lateral line often pink hued. Dorsal surface green to blue and covered in large, dark, oval spots that are visible from side and extend onto upper flank. Dorsal fin usually lacks spots. Anal fin leading edge shorter than base of fin. Adipose fin small, slender and often has an oval spot or two. Caudal fin lobes have large, oval blotches that are usually more muted than dorsal spots. Soft bodied. Slimy when handled. Mature individuals typically 35–55 cm (14–22 in) in length and 1–2.5 kg (2–5.5 lbs) in weight but can reach up to 6.5 kg (14 lbs). *Typical dead fish diagnostics: 27–33 long, slender rakers on first gill arch, 130–195 long, slender pyloric caeca, 14–16 anal fin rays, 11–14 branchiostegal rays.*

BEHAVIOUR AND HABITAT

Distributed offshore throughout the Gulf of Alaska, with some fish reaching the southern Bering Sea. Tends to stay within 30 m (100 ft) of the surface when offshore and within 20 m (65 ft) when migrating coastally back to fresh water. Does not change its depth as much between night and day as other species. Consumes a wide variety of prey, with little preference for any single prey. During its return migration, is commonly seen in large schools right at the surface, often leaping out of the water. In abundant years, schools can be enormous, with surface activity reaching as far as the eye can see. Often feeds around tidelines.

SIMILAR SPECIES

Not to be mistaken for a small Chinook, which has larger scales, a firmer body and wavy spots on the dorsal surface.

Chum Ocean Adult

A cantankerous and elusive fish with a dainty "wrist" and massive pupils that stare into your soul

KEY CHARACTERISTICS

Inside of mouth mostly white and pink. Lower jaw has grey or black gumline at base of teeth. Teeth relatively sharp and variable in size, sometimes absent but usually emerged by final approach to spawning river. Pupil large. Dorsal surface generally light or metallic blue and lacks spots, but sometimes has fine black speckling. Dorsal fin and caudal fin lack spots. Anal and pelvic fins often have prominent white tips. Anal fin leading edge shorter than base of fin. Adipose fin small and slender. Caudal peduncle narrow. Caudal fin has silver streaks that extend from base towards trailing edge and cover roughly half of fin. Caudal fin lacks rigidity and collapses easily when caudal peduncle is held. Mature individuals typically 60–85 cm (24–33 in) in length and 4–9 kg (9–20 lbs) in weight but can reach up to 17 kg (38 lbs). *Typical dead fish diagnostics: 20–26 short, blunt rakers on first gill arch, 160–185 pyloric caeca, 13–14 anal fin rays, 13–14 branchiostegal rays.*

BEHAVIOUR AND HABITAT

Lives almost exclusively offshore after its first summer at sea, mainly throughout the Gulf of Alaska but also the Bering Sea. While there is some spatial overlap between North American and Asian stocks of chum, most North American chum stay in the eastern North Pacific Ocean. Occupies the upper 25 m (80 ft) of surface waters at night, but during the day can reach depths of 60–120 m (200–400 ft). Less likely to make deeper excursions during cloudy days. Eats whatever is available. Stomach contents sometimes unidentifiable or gelatinous, suggesting a unique willingness (but not a preference) to consume jellyfish. Feeds extensively during its final year in the ocean, sometimes nearly doubling its weight, before migrating back to freshwater in schools.

SIMILAR SPECIES

The ocean-adult sockeye is similar but has smaller pupils, a thicker caudal peduncle, a dull grey caudal fin with no silver streaks and a white gumline at base of the teeth on the lower jaw.

Sockeye Ocean Adult

A sleek fish with exquisite scales and an absence of spots

KEY CHARACTERISTICS

Inside of narrow mouth white, pink, grey and black. Lower jaw grey or black with white gumline where teeth would be. Teeth usually absent, resulting in smooth gums, but can emerge close to spawning river (but still small and not sharp). Dorsal surface greenish blue to steel blue and lacks spots, but sometimes has fine black speckling. Dorsal fin absent of spots. Anal fin leading edge shorter than base of fin. Adipose fin small and slender. Caudal fin generally a dull uniform grey with no silver streaks or spots (but may rarely have a few spots on top quarter of upper lobe). Body often slimy. Mature individuals typically 45–65 cm (18–26 in) in length and 1.6–4.5 kg (3.5–10 lbs) in weight but can reach up to 17 kg (38 lbs). *Typical dead fish diagnostics: 32–37 long, slender, close-set rakers on first gill arch, 65–95 pyloric caeca, 14–16 anal fin rays, 13–15 branchiostegal rays.*

BEHAVIOUR AND HABITAT

A broadly distributed species whose oceanic range extends from the Gulf of Alaska into the western North Pacific Ocean and Bering Sea. Almost always moves offshore after its coastal migration as a post-smolt. Continuously moves while on the high seas, travelling thousands of kilometres each year and overlapping with other sockeye originating from rivers up and down the Pacific coast. Mostly a surface dweller, almost always residing in the upper 20 m (65 ft) at night and upper 40 m (130 ft) during the day. Has a diverse diet that is often composed mostly of zooplankton but eats some fish and will devour large quantities of squid when available. Often feeds around tidelines. Travels in moderate to large schools.

SIMILAR SPECIES

Although similar at first glance, an ocean-adult chum has a large pupil, a narrow caudal peduncle, silver streaks in the caudal fin and a grey or black gumline at the base of the teeth on the lower jaw.

Coho Ocean Adult

A blue bullet from above and bright silver from the side, willing to attack anything that invades its space

KEY CHARACTERISTICS

Inside of mouth mostly black and grey. Lower jaw has white or light grey gumline at base of needle-like teeth. Dorsal surface blue to black, occasionally greenish in coastal waters. Dorsal surface, dorsal fin and caudal fin occasionally absent of spots, but usually have small, dispersed spots that are variable in shape (though often circular). Anal fin leading edge shorter than base of fin. Adipose fin small, slender and occasionally spotted. Caudal fin spotting usually, but not always, isolated to upper lobe. Caudal fin rays feel rough, like edge of a dime (these are bone segments), when a fingernail is lightly run over them. Mature individuals typically 45–75 cm (18–30 in) in length (jacks can be as small as 25 cm or 10 in) and 2–7 kg (4.5–15 lbs) in weight (large individuals can reach up to 12 kg or 26 lbs). *Typical dead fish diagnostics: 19–27 short rakers on first gill arch, 50–85 pyloric caeca, 13–14 anal fin rays, 13–14 branchiostegal rays.*

BEHAVIOUR AND HABITAT

Mainly resides within the Gulf Alaska and mainly along the coast but many also move offshore after travelling along the continental shelf to the Aleutian Islands. Stays generally within 25 m (80 ft) of the surface on the high seas and shallower (upper 20 m or 65 ft) in coastal waters. Exhibits a strong tendency to occupy surface waters at night and move deeper during the day. A voracious predator that primarily feeds on fish, squid and krill, with the relative contributions of each varying from location to location. Often feeds extensively during its coastal migration back to its home river, causing bait balls of forage fish through which it darts in and out. Can move in groups or be relatively solitary throughout its ocean residency, but usually reconvenes to form tight schools when migrating back to freshwater.

SIMILAR SPECIES

Often mistaken for Chinook and occasionally for steelhead. Unlike coho, an ocean-adult Chinook has a black gumline at the base of the teeth on the lower jaw and caudal fin rays that feel smooth when a fingernail is lightly run along them. A Chinook also has a noticeable ridge on the upper caudal peduncle behind the attachment point of the outer caudal fin ray that is absent in coho. In contrast to coho, an ocean-adult steelhead has a white mouth on the inside, dense caudal fin spots that extend in rows parallel to the fin rays, a longer leading edge than base of the anal fin and a stark transition from dorsal colour to the white or silver of the flanks.

Chinook Ocean Adult

A darkly outlined giant that looks as if a black paintball exploded inside its mouth

KEY CHARACTERISTICS

Inside of mouth almost entirely black. Lower jaw has black gumline at base of sharp, conical teeth. Dorsal surface green to black. Dorsal surface, dorsal fin and caudal fin covered in medium-sized and moderately dense black spots that become more pronounced during final approach to spawning river. Dorsal surface has irregularly shaped (wavy) spots; fin spots are usually more circular. Anal fin leading edge shorter than base of fin. Adipose fin large, stout and occasionally spotted. Caudal fin lobes have spotting usually, but not always, present on both lobes. Caudal fin ray feels smooth (no bone segments) when a fingernail is lightly run over them; outer rays stiff and do not collapse when caudal peduncle is held. Deep bodied. Often smells metallic, musky fishy or (some say) like pungent kelp. Noticeably dark outline and features when underwater. Mature individuals typically 55–115 cm (22–45 in) in length (jacks can be as small as 25 cm or 10 in) and 4.5–20 kg (10–44 lbs) in weight (large individuals can reach up to 57 kg or 126 lbs). *Typical dead fish diagnostics: 20–25 short rakers on first gill arch, 140–185 pyloric caeca, 15–19 anal fin rays, 16–18 branchiostegal rays.*

BEHAVIOUR AND HABITAT

The deepest-dwelling salmon species, but also one of the most coastal. Usually stays within a couple hundred kilometres of the coastline, but occasionally ventures off the continental shelf, especially in the Bering Sea compared to the Gulf of Alaska. Generally found 20–50 m (65–165 ft) in depth when offshore and 10–150 m (30–500 ft) in depth in coastal oceans. When moving through shallow (<50 m or 165 ft) water, often found near the ocean floor around featured terrain. Larger fish occasionally caught in depths down to 400 m (1300 ft). Moves in shallower water and at greater speeds at night than in the day. Eats most large prey items (up to 40% of its body length), but preferentially feeds on fish, with most of its diet usually composed of forage fishes like sardine, anchovy and herring. More individualistic than most other species, but still forms loose schools.

SIMILAR SPECIES

Commonly confused with coho, but the ocean-adult coho has a white or light grey gumline at the base of the teeth on the lower jaw and caudal fin rays that feel rough (like the edge of dime) when lightly running a fingernail along them. Coho also lacks the Chinook ridge on the upper caudal peduncle behind the attachment point of the outer caudal fin ray. Smaller Chinook can also be mistaken with adult pinks, but pinks have small scales, soft bodies and large oval spots on the dorsal surface and caudal fin.

Steelhead Ocean Adult

A distinctly trout-like fish that sports a short, rounded snout and rows of small caudal fin spots that look as if they were dotted on with a paintbrush

KEY CHARACTERISTICS

Snout from side is short and rounded. Inside of mouth almost entirely white. Lower jaw has a white gumline at base of sharp teeth. Maxillary may or may not extend past rear margin of eye (unlike earlier steelhead stages). Dorsal surface colour metallic blue to dark blue. Dorsal colour transitions immediately to white or silver on flank, rather than fading gradually. Spotting on upper flank mostly above lateral line. Dorsal surface, dorsal fin and caudal fin covered in small, dense circular black spots. Dorsal fin and caudal fin spots aligned in rows extending parallel to fin rays on both lobes of caudal fin. Anal fin leading edge shorter than base of fin. Adipose fin typically spotted with a small base. Caudal fin squarish, often flat along trailing edge with outer rays extending less steeply from body than in other species. Mature individuals typically 45–90 cm (18–35 in) in length and 1.8–7 kg (4–15 lbs) in weight but can reach up to 17 kg (37 lbs). Intermediate characteristics may indicate a cutthroat-rainbow hybrid, but genetic tools are needed to confirm. *Typical dead fish diagnostics: 17–21 short, blunt rakers on first gill arch, 30–70 pyloric caeca, 8–12 anal fin rays, 10–15 branchiostegal rays, no basibranchial teeth on base of tongue.*

BEHAVIOUR AND HABITAT

The widest ranging and most southerly of all seven species, spanning most of the North Pacific Ocean from east to west, although not up into the Bering Sea. While some stay on the continental shelf throughout their ocean residency, most move offshore, often migrating immediately as post-smolts rather than travelling northwards along the coast. Depth preference is mostly unknown but tends to stay within 5 m (16 ft) of the surface during the final approach to its home river. Mainly feeds on fish, krill and, when available, squid. Schooling behaviour when offshore is unclear but is mostly solitary when coastal.

SIMILAR SPECIES

Unlike steelhead, an ocean-adult coho has a black mouth on the inside, a shorter leading edge than base of the anal fin and far fewer caudal fin spots that do not extend in rows parallel to the fin rays. The transition of colour from the dorsal surface towards the lateral line is also less stark in coho than in steelhead.

Coastal Cutthroat Ocean Adult

Nearly tropically colourful and generously spotted

KEY CHARACTERISTICS

Tip of snout often yellow around mouth. Inside of mouth almost entirely white. Lower jaw has white gumline at base of sharp teeth. Facial spotting often present below eye and sometimes on operculum. Dorsal surface colour dependent on habitat (bluer in the ocean, yellowish green or brown at beaches or estuaries). Flank spotting extends below lateral line. Flank, dorsal surface, dorsal fin and caudal fin covered in small, dense, circular black spots. Anal fin leading edge shorter than base of fin. Adipose fin small, slender and usually spotted. Caudal fin spots aligned in rows extending parallel to fin rays on both lobes of caudal fin. All fins at least partially yellow, orange or brown. Typically 30–50 cm (12–20 in) in length and 1–2 kg (2–4.5 lbs) in weight. Occasionally has diagnostic faint yellow, orange or red slashes under lower jaw. Intermediate characteristics may indicate a cutthroat-rainbow hybrid, but genetic tools are needed to confirm. *Typical dead fish diagnostics: 15–21 short, blunt rakers on first gill arch, 25–55 pyloric caeca, 8–12 anal fin rays, 9–12 branchiostegal rays, basibranchial teeth present on base of tongue.*

BEHAVIOUR AND HABITAT

Mostly lives in or around shallow nearshore beaches with eelgrass or other nursery features, much like it did as a post-smolt. Usually stays in its natal estuary or migrates among local estuaries, but some populations in Washington, Oregon and California have been caught up to 70 km (45 mi) offshore (though still on the continental shelf) in spring and early summer. British Columbian and Alaskan populations seem to remain in protected inlets along the coast rather than moving offshore. Feeds almost exclusively on fish. Ambushes schools of smaller fish foraging near the shore so energetically that it can accidently beach itself. Forms small, loose groups when nearshore, but schooling behaviour off the coast is unclear.

SIMILAR SPECIES

The ocean-adult coastal cutthroat is quite distinct, with yellow or orange on the fins and dense spotting extending well below the lateral line and onto the face.

SPAWNERS

The spawning life stage of the salmon life cycle is one of the most unique exhibitions in the animal kingdom. Every year, hundreds of millions of Pacific salmon migrate up their spawning rivers to reproduce and die (except for steelhead and coastal cutthroat, which can return to spawn multiple times). People are generally more familiar and intrigued with this life stage than any other, with its large aggregations of charismatic, brightly coloured fish that, over just a few weeks, transform into a mass die-off. Through their carcasses, Pacific salmon move billions of tonnes of nutrients from the sea into the land and rivers of western North America.

For many salmon, the ocean migration to the mouth of rivers is only the beginning of their journey to reproduce, as some populations must still migrate up to 3200 km (2000 mi) to reach their spawning grounds. The arduous upriver migration for these distant inland populations is now more perilous than ever because of dams, fishing and warming water temperatures.

Salmon will often hold outside the mouths of rivers, acclimatizing to freshwater and waiting for rain to raise water levels and lower water temperatures to make their upcoming journey easier. In big rivers like the Fraser, these "staging" salmon can number in the millions, slowly growing in numbers each sunny day and causing a flurry of activity as seals, sea lions and humans all try to catch as many fish as possible. Once conditions are right, salmon usually begin moving upriver on a slack or flooding (incoming) tide to limit their energy expenditure. This timing of travel with the tides can continue for many days and occur much farther upriver than one might think—for example, the Bonneville Dam is 235 km (146 mi) from the mouth of the Columbia River but still experiences tidal influence.

The return to freshwater of the five species that die after spawning marks the point at which they stop eating and instead live off the energy reserves they built up in the ocean. Some salmon will still occasionally feed in the river, especially on salmon eggs, but this is normally an opportunistic rather than deliberate behaviour.

Except for steelhead and coastal cutthroat, which are adapted to multiple trips between river and ocean, the more time adult salmon spend in freshwater the more their bodies shut down. They become progressively skinnier as they use up their energy reserves, and their organs and immune systems stop working well. Soon after spawning (and sometimes before), salmon begin to visibly decay while still alive, losing skin and pieces of their bodies and gaining open wounds and fungal infections. At this point they are often called "zombie fish" because they look like dead fish swimming. Here, we present salmon in full spawning colours with no visible decay.

Female salmon prepare their redds by digging with their caudal fins to create a depression into which they will eventually deposit their eggs. Prime locations for redds are heavily sought after, so females are constantly guarding them. Males, on the other hand, compete with other males for access to females that are ready to spawn. Competition often starts with posturing for dominance but, when that is not sufficient, ends in chasing, ramming and biting.

To mate, the male moves from one side of the female to the other multiple times and then stops to quiver next to her. The female releases a portion of her eggs through her vent and the male releases his milt (seminal fluid) as close to the female's vent as possible. Sometimes other males will linger just behind a mating pair and try to sneak in to fertilize the eggs; these are called sneaker males and are almost always smaller than the dominant males. Sneaker males are usually jacks or, in the case of steelhead, small freshwater resident trout.

After mating, the female gently covers the fertilized eggs with gravel, while at the same time forming a second pocket for another spawning event. She will repeat this multiple times, sometimes with the same male and sometimes with another, until all her eggs are depleted. Females of the five salmon species that die after spawning continue guarding their redds until they are too weak to hold their position in the stream, after which they drift downstream and die. Female steelhead and coastal cutthroat usually leave their redds after spawning but may return over the next day or two to cover the redds with more gravel.

Some adult salmon begin gaining their spawning colours during their ocean migration towards their spawning river, whereas others are still silver in the river until their final approach to the spawning grounds. The timing of this colour transition varies greatly among species, populations and individuals. Pink and chum often gain faint spawning colours and patterning while still in the ocean hundreds of kilometres from their rivers, whereas Chinook and coho are sometimes still silver after travelling hundreds of kilometres upriver. The same holds true for other spawning features like the hooked jaw and dorsal hump in males of some species.

Spawning salmon arrive at predictable times and places with most spawning in the late summer to mid-fall but steelhead, coastal cutthroat and some Chinook spawn in the late winter or spring. Once spawning starts in earnest, you can usually identify a spawning reach of a salmon river by the smell of dead fish. Look closer and you will probably find carcasses along the shoreline, slow-moving individuals in the shallows and the occasional leaping fish in deeper water. If you don't see or smell dead fish, look for stretches of river with gravel or cobble bottoms (high-quality spawning habitat) rather than mud or other fine sediment.

Spawning salmon are as easy as it gets for salmon identification. Most species are totally unique, but Chinook and coho can still be mildly troublesome for novice observers. Unlike the other stages in the salmon life cycle, distinguishing male and female spawners is straightforward, so we show illustrations for both.

SUMMARY TABLE OF SPAWNERS KEY CHARACTERISTICS

Feature	Pink	Chum	Sockeye
Female colour and patterning	olive green overall; white blotching below lateral line; faded pink stripe along lateral line	dusky green overall; "wine-stained" flanks (vertical bars of dirty red or purple)	grey-red flank and dorsal and adipose fins; pale green caudal fin and head (except lower jaw)
Male colour and patterning	similar to females but darker grey-green overall; fins dark; lateral line red	same as females but red colouration is more pronounced	typically brighter and more extensively red than females; typically black on snout, maxillary and operculum
Other sexual dimorphism	males have elongated snout, hooked upper jaw, slightly hooked lower jaw, enlarged teeth, prominent hump	males have elongated snout, hooked jaws, enlarged teeth	males have elongated snout, hooked jaws, enlarged teeth, hump behind the head
Caudal fin spotting	large, dark, oval blotches on both lobes	none	typically none, but can gain fine black spotting on top quarter of upper lobe or along trailing edge of fin
Dorsal spotting	large, dark, oval spots	none, but sometimes fine black speckling	none, but sometimes fine black speckling
Typical weight	1–2.5 kg (2–5.5 lbs)	4–9 kg (9–20 lbs)	1.6–4.5 kg (3.5–10 lbs)
Typical length	35–55 cm (14–22 in)	60–85 cm (24–33 in)	45–65 cm (18–26 in)
Other key features	scales small and loose	caudal peduncle narrow; pupil large; anal and pelvic fins with prominent white tips	
Most similar species	unique	unique	unique

Coho	Chinook	Steelhead	Coastal Cutthroat
variable; flank dark with bronze or pinkish red patch or patches; head green-black; operculum pinkish red	dark green or green-black overall, pinkish red area that fades into darkness towards head and dorsal surface	light brown or green-brown overall, operculum rosy, continuous band of rosy red along lateral line and sometimes lower flank	golden yellow, yellow-brown or dull bronze overall; operculum rosy red; belly and/or lateral line sometimes rosy
flank can be brilliant bronze-red throughout or a muted red gradient that darkens towards dorsal surface; head green-black; operculum pinkish red	same as females except red area is brighter and more extensive when present	same as females except lateral line and operculum are usually redder	same as females but often rosier overall
males have elongated snout, deeply hooked upper jaw, hooked lower jaw, enlarged teeth	males have elongated snout, slightly hooked jaws, enlarged teeth	males have elongated snout, often a vertical protrusion at end of lower jaw	males have elongated snout
sometimes none, but when present are small, dispersed, circular black spots, usually only on upper lobe	large, moderately dense, circular black spots, usually on both lobes but not always; most variable of all species	small, dense, circular black spots on both lobes, extending in rows parallel to fin rays	medium-sized, dense, circular black spots on both lobes, extending in rows parallel to fin rays
small circular or irregular black spots	large, moderately dense, irregular (wavy) black spots; most variable of all species	small, dense, circular black spots, mostly above lateral line	medium-sized, dense, circular black spots, often extending below lateral line
2–7 kg (4.5–15 lbs)	4.5–20 kg (10–44 lbs)	1.8–7 kg (4–15 lbs)	1–2 kg (2–4.5 lbs)
45–75 cm (18–30 in)	55–115 cm (22–45 in)	45–90 cm (18–35 in)	30–50 cm (12–20 in)
fin rays of caudal fin feel rough (like edge of dime) when lightly running fingernail along them (unlike Chinook); inside of mouth white or grey; gumline at base of teeth on lower jaw black or grey; nostril outlined white	fin rays of caudal fin feel smooth when lightly running fingernail along them (unlike coho); inside of mouth black; gumline at base of teeth on lower jaw black	caudal fin has flat trailing edge	spots often on and/or below eye
Chinook or steelhead	coho	coho	unique

Pink Spawner

A small but exuberant fish that can overwhelm streams with its numbers

KEY CHARACTERISTICS

Snout elongated in males. In Alaska, female pinks also occasionally have elongated snouts (as in illustration). Inside of mouth white, pink, grey and black. Upper jaw hooked; lower jaw slightly hooked in males; lower jaw has black gumline at base of small teeth. Teeth enlarged in males. Flank olive green (females) or dark grey-green (males) that is lighter and with large white blotches below lateral line. Lateral line has pink (females) or red (males) band that is thicker near caudal fin. Scales small and loose. Prominent hump behind head of males. Dorsal surface covered in large, dark, oval spots visible from side and extending onto upper flank. Fins olive green (females) or dark grey with highlights of variable colour (males). Dorsal fin usually lacks spots. Adipose fin small and slender and often has oval spots. Caudal fin lobes have large, oval blotches that are usually more muted than dorsal spots. Soft bodied. Typically 35–55 cm (14–22 in) in length and 1–2.5 kg (2–5.5 lbs) in weight but can reach up to 6.5 kg (14 lbs). *Typical dead fish diagnostics: 27–33 long, slender rakers on first gill arch, 130–195 long, slender pyloric caeca, 14–16 anal fin rays, 11–14 branchiostegal rays.*

BEHAVIOUR AND HABITAT

Spawning occurs in late summer or early fall in lower reaches of small coastal streams or tributaries of larger rivers, sometimes even in tidally influenced estuary habitat. Forms redds in small gravel (on average 9 mm or 0.4 in). Larger males have greater access to females, but they may not be more reproductively successful if female ripeness or male arrival to the spawning grounds is mistimed. Males with larger humps may have a competitive advantage over those with smaller humps.

SIMILAR SPECIES

The pink spawner is unique with a combination of small body size, oval spotting and prominent male dorsal hump.

Chum Spawner

A big brute with unmistakable wine stains down its sides

KEY CHARACTERISTICS

Snout elongated in males. Inside of mouth mostly white and pink. Jaws hooked in males; lower jaw has grey or black gumline at base of teeth. Teeth enlarged in males. Pupil large. Flank dusky green overall with vertical bars of dirty red or purple that look like wine stains or paint running down side of fish. Females and subordinate males sometimes have a thick horizontal band of reddish purple along lateral line rather than vertical bars. Red colouration more pronounced in males than females. Dorsal surface lacks spots, but sometimes has fine black speckling. Dorsal and caudal fins lack spots. All fins dusky or dark green. Anal and pelvic fins often have prominent white tips. Adipose fin small and slender. Caudal peduncle narrow. Caudal fin lacks rigidity and collapses easily when caudal peduncle is held. Typically 60–85 cm (24–33 in) in length and 4–9 kg (9–20 lbs) in weight but can reach up to 17 kg (37 lbs). *Typical dead fish diagnostics: 20–26 short, blunt rakers on first gill arch, 160–185 pyloric caeca, 13–14 anal fin rays, 13–14 branchiostegal rays.*

BEHAVIOUR AND HABITAT

Spawning occurs in mid-fall (except for the unique summer runs in the Yukon and Amur rivers) in lower reaches of small coastal streams or tributaries of larger rivers. Forms redds in most gravel sizes (on average 25 m or 1 in). Females delay redd preparation when males smaller than themselves are present; males appear to choose females based on their readiness to spawn (redd preparation) rather than their size. Larger males dominate competition for females through posturing and combat.

SIMILAR SPECIES

Vertical bands of red and purple on the flank distinguish the chum spawner from all other species.

Sockeye Spawner

The most vibrantly coloured and easily recognizable of all salmon

KEY CHARACTERISTICS

Snout elongated in males. Head pale green with dark green or black snout, maxillary and lower operculum (especially in males). Inside of mouth white, pink, grey and black. Jaws hooked in males. Lower jaw grey or black with white gumline at base of teeth. Teeth enlarged in males. Hump behind head of males. Flank grey-red (females) or bright red (males). Dorsal surface lacks spots, but sometimes has fine black speckling. Dorsal fin lacks spots. Dorsal and adipose fins grey-red (females) or bright red (males). Adipose fin small and slender. Pectoral, pelvic and anal fins green, grey or black, sometimes with red near flank (especially in males). Caudal fin pale green, sometimes with red near caudal peduncle (especially in males). Caudal fin usually lacks spots but can gain fine black spotting on top quarter of upper lobe or along trailing edge of fin. Typically 45–65 cm (18–26 in) in length and 1.6–4.5 kg (3.5–10 lbs) in weight but can reach up to 17 kg (37 lbs). *Typical dead fish diagnostics: 32–37 long, slender, close-set rakers on first gill arch, 65–95 pyloric caeca, 14–16 anal fin rays, 13–15 branchiostegal rays.*

BEHAVIOUR AND HABITAT

Spawning occurs from early summer to late fall, with timing strongly dependent on the population, in lakes or reaches of rivers close to lakes so that the emerging fry do not have to travel far to rear. Forms redds in highly variable gravel sizes, from sandy lake beaches to rocky lake shores with rocks up to a third of their body length. Males try to court large females, which produce more eggs, and females with a specific wavelength and brightness of red. Dominant males, which get the most mating opportunities, tend to be those with longer bodies, longer snouts and larger humps.

SIMILAR SPECIES

With a bright red body and green head and tail, the sockeye spawner is as unique as salmon get.

Coho Spawner

A spirited fish that could swim its way up a wet rope to reach its spawning grounds

KEY CHARACTERISTICS

Snout elongated in males. Inside of mouth mostly black and grey. Upper jaw deeply hooked and lower jaw hooked in males. Lower jaw has white or light grey gumline at base of needle-like teeth. Teeth enlarged in males. Nostrils outlined white. Head dark green to black with operculum pinkish red. Flank colour variable: dark flank with one or more bronze or pinkish red patches (females); brilliant bronze-red to muted red gradient that darkens towards dorsal surface (males). Dorsal surface and dorsal fin have small spots that are variable in shape (though often circular). Fins variable in colour, including dark green, bronze-red or black. Adipose fin small, slender and often spotted. Caudal fin spots (when present) are small, circular and usually, but not always, isolated to upper lobe. Fin rays of caudal fin feel rough, like edge of a dime (these are bone segments), when a fingernail is lightly run over them. Typically 45–75 cm (18–30 in) in length (jacks can be as small as 25 cm or 10 in) and 2–7 kg (4.5–15 lbs) in weight (large individuals can reach up to 12 kg (26 lbs). *Typical dead fish diagnostics: 19–27 short rakers on first gill arch, 50–85 pyloric caeca, 13–14 anal fin rays, 13–14 branchiostegal rays.*

BEHAVIOUR AND HABITAT

Spawning occurs between mid-summer and late fall, in rivers throughout the watershed but most commonly in small coastal streams. Forms redds in medium-sized gravel (on average 18 mm or 0.7 in). Larger females are more successful at defending their redds against other females and at obtaining high-quality redd sites that result in higher egg survival. Males tend to be quite aggressive towards other males, and this is rewarded with greater access to females. Wild coho males are more aggressive than those raised in a hatchery and therefore have higher reproductive success.

SIMILAR SPECIES

Occasionally confused for Chinook or steelhead. Unlike coho, the Chinook spawner has a fully black mouth with a black gumline at the base of teeth of the lower jaw and caudal fin rays that feel smooth when lightly running a fingernail along them. A steelhead spawner has a white mouth on the inside, reddish colour along the lateral lines and many caudal fin spots arranged in rows parallel to the fin rays.

Chinook Spawner

A behemoth that can take your breath away with its size and strength yet whose calm demeaner belies its battle scars

KEY CHARACTERISTICS

Snout elongated in males. Inside of mouth almost entirely black. Jaw slightly hooked in males. Lower jaw with black gumline at base of sharp, conical teeth. Teeth enlarged in males. Flank either dark green throughout (see appendix for illustrations of dark-green Chinook spawners) or green-black with a reddish area (individuals within a population will typically all be one or the other) that is generally pinker and smaller in females than in males and tends to fade into darkness towards the head and dorsal surface. Dorsal surface, dorsal fin and caudal fin covered in large and moderately dense black spots that become more pronounced during the final approach to the spawning river. Dorsal surface spots irregularly shaped (wavy); fin spots usually more circular. Adipose fin large, stout and occasionally spotted. Caudal fin lobes usually, but not always, both spotted. Caudal fin rays feel smooth (no bone segments) when a fingernail is lightly run over them. Outer caudal fin rays stiff and do not collapse when caudal peduncle held. Deep bodied. Typically 55–115 cm (22–45 in) in length (jacks can be as small as 25 cm or 10 in) and 4.5–20 kg (10–44 lbs) in weight (large individuals can reach up to 57 kg or 126 lbs). *Typical dead fish diagnostics: usually 20–25 short rakers on first gill arch, 140–185 pyloric caeca, 15–19 anal fin rays, 16–18 branchiostegal rays.*

BEHAVIOUR AND HABITAT

Spawning can occur in any season, with timing dependent on the population and life history. Spawns in the mainstem or tributaries of large rivers. Forms redds in large gravel (on average 35 mm or 1.4 in) with minimal fine sediment, which markedly decreases egg survival. Females prefer large males and will act aggressively towards small males and delay spawning for several hours in their presence. Larger females obtain the best redd sites with optimal gravel size and depth. White-fleshed Chinook do not exhibit a preference for white-fleshed mates over red-fleshed mates (nor vice versa).

SIMILAR SPECIES

Coho can be similar but does not have the fully black mouth with black gumline at the base of the teeth on the lower jaw, and caudal fin rays are rough when you run a fingernail over them. The Chinook spawner also lacks the white-outlined nostrils and deeply hooked upper jaw of the coho spawner.

♀

♂

Steelhead Spawner

A beautiful and increasingly rare fish with power to spare

KEY CHARACTERISTICS

Snout elongated in males. Inside of mouth almost entirely white. Lower jaw in males often has a vertical protrusion on end. Lower jaw has white gumline at base of sharp teeth. Operculum rosy or red (usually redder in males). Flank light brown or green-brown, with a continuous band of pink or red along lateral line and sometimes lower flank (usually redder in males). Upper flank spotting mostly above lateral line. Dorsal surface, dorsal fin and caudal fin covered in small, dense circular black spots. Dorsal fin and both lobes of caudal fins have spots aligned in rows extending parallel to fin rays. Dorsal and caudal fins roughly same colour as caudal peduncle; other fins can be lighter or darker. Adipose fin typically spotted with a small base. Caudal fin squarish, often flat along trailing edge. Typically 45–90 cm (18–35 in) in length and 1.8–7 kg (4–15 lbs) in weight but can reach up to 17 kg (37 lbs). Females vary in depth and fullness, depending on whether they've spawned (illustration is post-spawn). Intermediate characteristics may indicate a cutthroat-rainbow hybrid, but genetic tools are needed to confirm. *Typical dead fish diagnostics: 17–21 short, blunt rakers on first gill arch, 30–70 pyloric caeca, 8–12 anal fin rays, 10–15 branchiostegal rays, no basibranchial teeth on base of tongue.*

BEHAVIOUR AND HABITAT

Spawning occurs in spring. Steelhead may enter the river in winter or spring (as sexually mature individuals that spawn within a month or two) or enter in summer or fall (maturing while in the river until spring). Typically spawns in medium to large coastal rivers but may migrate inland. Forms redds in medium-large gravel (average 25 mm or 1 in) with minimal fine sediment, which markedly decreases egg survival. Usually mates with other steelhead but females can mate with resident rainbow trout as well, often via small rainbow sneaker males that fertilize eggs when a steelhead pair is spawning. Mate choice is determined more by arrival date to the spawning grounds than by body size.

SIMILAR SPECIES

Coho is similar but can be distinguished by its black mouth (on the inside), few caudal fin spots that do not extend in rows parallel to the fin rays, white-outlined nostrils and deeply hooked upper jaw.

Coastal Cutthroat Spawner

An elegant fish forsaking gaudy colour for tasteful tones and a liberal spattering of spots

KEY CHARACTERISTICS

Snout elongated in males. Tip of snout often yellow around mouth. Inside of mouth almost entirely white; lower jaw has white gumline at base of sharp teeth. Spotting on face often present below eye and sometimes on operculum. Operculum rose tinted to rosy red, with females tending towards the former and males towards the latter. Diagnostic faint red, orange or rarely yellow slashes under lower jaw. Flank golden yellow, yellow-brown or dull bronze, often with a pale orange or rosy belly and/or lateral line, especially in males. Males generally darker and more colourful overall. Flank, dorsal surface, dorsal fin and caudal fin covered in small, dense, circular black spots. Dorsal and caudal fins roughly same colour as caudal peduncle; other fins can be lighter or darker. Adipose fin small, slender and usually spotted. Caudal fin spots aligned in rows extending parallel to fin rays on both lobes of caudal fin. The anadromous form is typically 30–50 cm (12–20 in) in length and 1–2 kg (2–4.5 lbs) in weight but lake residents are often larger (achieving sizer greater than 4.5 kg or 10 lbs) and stream residents are smaller (generally less than 0.2 kg or 0.5 lbs). Females vary in depth and fullness, depending on whether they've spawned. Intermediate characteristics may indicate a cutthroat-rainbow hybrid, but genetic tools are needed to confirm. *Typical dead fish diagnostics: 15–21 short, blunt rakers on first gill arch, 25–55 pyloric caeca, 8–12 anal fin rays, 9–12 branchiostegal rays, basibranchial teeth present on base of tongue.*

BEHAVIOUR AND HABITAT

Spawning occurs over a protracted window in winter and spring, generally in small tributaries and headwaters of coastal streams with low gradients and flows. Forms redds in pea-sized gravel (9 mm or 0.4 in). Spawning typically occurs in riffles 15–45 cm (6–18 in) deep, often near deep pools for potential cover. Males aggressively with gaping jaws pursue other males (and other fish species) and will often guard redds while females cover the eggs. Commonly mates with other anadromous individuals but can mate with coastal cutthroat of other life histories as well, including small male freshwater residents that employ sneaker tactics to fertilize the eggs of an actively spawning anadromous female. Little is known about mate choice.

SIMILAR SPECIES

No other species has throat slashes or spots that extend well below the lateral line and onto the face.

GLOSSARY

alevin: a newly hatched salmon that lives off its nutrient-rich yolk sac for several weeks or months in the interstitial spaces of its redd

anadromous: fish, such as salmon, that are born in freshwater, mature in the ocean and return to freshwater to spawn

basibranchial teeth: the small teeth on the basibranchial bone at the base of the tongue

benthic: associated with the bottom of a body of water

branchiostegal rays: the thin bones just beneath and behind the operculum

caudal peduncle: the narrow part of the body of a fish between the anal and caudal fins; often called the "wrist"

coastal: the oceanic waters on the continental shelf

dimpling: the raindrop-like pattern on the water's surface that is caused by the movement of a group of fish

extirpated: locally (but not globally) extinct

food web: the collection of energy-transfer relationships among organisms in an ecosystem (such as via predation or herbivory)

fork length: the distance from the tip of the snout to the fork of the tail

fry: a juvenile salmon that has absorbed its yolk sac and emerged from the gravel

gill rakers: the filaments on the inside of a gill arch

gravel emergence: the time at which juvenile salmon rise above the interstitial spaces of their redds to inhabit the water column, marking the transition from alevin to fry

glide: a section of creek just below a pool

headwaters: the uppermost sections of watersheds, composed of small creeks and streams

jack: a male salmon that matures young, migrating back to spawn one to three years early at small body size and without spawning colouration, and which opportunistically sneaks in to fertilize the eggs of a female that is spawning with a larger male

kelt: a repeat-spawning fish, such as steelhead or coastal cutthroat, that is in the process of returning to the ocean

life history: the patterns of growth, survival, migration, and reproduction that describe the life cycle of a species

nearshore: within a short distance of the shore

ocean adult: a mature salmon in the marine environment, either coastal or offshore

offshore: the oceanic waters beyond the continental shelf

otoliths: the calcified structures in the inner ear of bony fishes that can be used to age fish by counting and measuring their incremental depositions, like counting tree rings

outmigration: the juvenile migration of salmon to the open ocean, from rivers to estuaries and from coastal waters to offshore feeding grounds

operculum: the bony flap that protects and covers the gills

parr: a juvenile salmon that remains in freshwater for more than one year

parr marks: the series of dark vertical markings on the flanks of many juvenile salmon

pelagic: associated with the water column (rather than the bottom)

phytoplankton: small, mostly microscopic, photosynthetic organisms that form the base of aquatic food webs

pool: a section of stream that is deeper and slower moving than nearby sections

post-smolt: a juvenile salmon in its early marine life, here defined as a salmon in the ocean between 10 cm (4 in) and 25 cm (10 in) in fork length

pyloric caeca: the thin tube-like appendages attached to the digestive tract at the interface of the stomach and intestines in many fishes, including salmon

redd: a nest made of gravel in which salmon deposit their eggs

riffle: a shallow section of river or stream with fast-flowing water

run: the total number of adult salmon in a stream or river migrating to spawn

school: a group of fish swimming together in the same direction

shoal: a group of fish swimming together but not in the same direction

seine: a long fishing net that hangs like a fence from a floating line at the surface and is used to encircle and catch fish

sexual dimorphism: the differences in non-reproductive characteristics between males and females of the same species

shore: the interface of land and water

smolt: a juvenile salmon transitioning from freshwater to saltwater during or just after its migration downriver

smoltification: the process through which juvenile salmon transition from freshwater to marine form via changes to their morphology, physiology and behaviour.

sneaker male: a subdominant spawning male that lingers just behind a mating pair and tries to surreptitiously and opportunistically fertilize the eggs; are usually jacks but in the case of coastal cutthroat and steelhead are typically small freshwater resident trout

spawner: an adult salmon in freshwater just before, during or after reproduction; bearing distinct changes in colouration and morphology compared to its oceanic form

unzipped: the phase during which a salmon fry has not fully absorbed its yolk sac and still has a ventral slit

watershed: the area of land that drains surface water and groundwater into a common waterbody, such as a creek, river, lake or ocean

yolk sac: the external pouch containing nutrient-rich yolk that salmon retain on their abdomen for nourishment during their first few weeks after hatching

zooplankton: small, mostly microscopic aquatic animals, such as copepods, pteropods, euphausiids (krill) and amphipods, that feed on phytoplankton and other zooplankton

APPENDIX

Dark-green Chinook spawners

KEY REFERENCES

We leaned on the work of those before us to supplement our own knowledge when writing this field guide. Here is a selection of resources that were particularly important to this work. We encourage readers to explore these for further information.

Augerot, W. *Atlas of Pacific Salmon*. Berkeley: University of California Press, 2005.

Behnke, J. *Trout and Salmon of North America*. New York: The Free Press, 2002.

Groot, C. and L. Margolis. *Pacific Salmon Life Histories*. Vancouver: UBC Press, 1991.

Pearcy, W.G., R.D. Brodeur, S.M. McKinnell and J.P. Losee. Ocean ecology of anadromous coastal cutthroat trout. In R. Beamish. Ed. *The Ocean Ecology of Pacific Salmon and Trout*. Bethesda: American Fisheries Society, 2018.

Pollard, W., G. Hartman and C. Groot. *Field Identification of Coastal Juvenile Salmonids*. Madeira Park: Harbour Publishing Company, 1997.

Quinn, T.P. *The Behavior and Ecology of Pacific Salmon and Trout* (Second ed.). Vancouver: UBC Press, 2018.

Shaw, W. Biological sampling manual for salmonids: a standardized approach for the Pacific Region. *Canadian Technical Report of Fisheries and Aquatic Sciences*, 1998.

Trautman, M.B. *A Guide to the Collection and Identification of Presmolt Pacific Salmon in Alaska with an Illustrated Key*. Seattle: National Oceanic and Atmospheric Administration, 1973.

ABOUT THE AUTHORS

Dr. Sean Godwin is a marine ecologist whose love for salmon was nurtured at an early age by his grandfather on a small boat off the coast of southern Vancouver Island. After a brief foray into biochemistry as an undergraduate, Sean returned to his roots and fell back in love with the coastal waters of the Pacific Northwest and the extraordinary fishes that live there. He completed his PhD at Simon Fraser University in 2018 and is currently a postdoctoral fellow at Dalhousie University, where he studies the early marine life of juvenile Pacific salmon. Sean also serves as a member of the marine fishes subcommittee of COSEWIC—the independent committee that assesses the status of endangered wildlife in Canada—and consults for the Mamalilikulla, 'Namgis and Kwikwasut'inuxw Haxwa'mis First Nations as they transition salmon farms out of their traditional territories. Sean lives in Vancouver with his adventure dog, Cashew, and can often be seen climbing (or, more likely, falling from) cliffs in Squamish.

Dr. Martin Krkosek is an Associate Professor and Canada Research Chair in the Department of Ecology and Evolutionary Biology at the University of Toronto, where he leads a research program in marine ecology and infectious disease. He has worked on Pacific salmon for most of his career, including over two decades of fieldwork on the marine life of juvenile salmon in the coastal waters of Kwakwaka'wakw territories between Vancouver Island and the British Columbia mainland. Much of his team's fieldwork has been based at the Salmon Coast Field Station, on Gilford Island. He has published over 80 scientific articles and one edited book in the areas of fisheries, conservation and epidemiology. When not teaching about ecology and mathematical modelling at the historic St. George campus of U of T, he can often be found with his students and young family mucking around with boats and fish in coastal BC.